SCIENCE
A CLOSER LOOK

BUILDING SKILLS

Reading and Writing Workbook

Contents

LIFE SCIENCE

Unit A Literature 1

Chapter 1 Cells and Kingdoms
Chapter Concept Map.. 2
- **Lesson 1** Lesson Outline 3
  - Lesson Vocabulary 5
  - Lesson Cloze Activity 6
- **Lesson 2** Lesson Outline 7
  - Lesson Vocabulary 9
  - Lesson Cloze Activity 10
  - Reading in Science 11
- **Lesson 3** Lesson Outline 13
  - Lesson Vocabulary 15
  - Lesson Cloze Activity 16
  - Writing in Science 17
- **Lesson 4** Lesson Outline 19
  - Lesson Vocabulary 21
  - Lesson Cloze Activity 22
  - Writing in Science 23
- **Lesson 5** Lesson Outline 25
  - Lesson Vocabulary 27
  - Lesson Cloze Activity 28

Chapter Vocabulary 29

Chapter 2 Parents and Offspring
Chapter Concept Map... 31
- **Lesson 1** Lesson Outline 32
  - Lesson Vocabulary 34
  - Lesson Cloze Activity 35
  - Writing in Science 36
- **Lesson 2** Lesson Outline 38
  - Lesson Vocabulary 40
  - Lesson Cloze Activity 41
- **Lesson 3** Lesson Outline 42
  - Lesson Vocabulary 44
  - Lesson Cloze Activity 45
- **Lesson 4** Lesson Outline 46
  - Lesson Vocabulary 48
  - Lesson Cloze Activity 49
  - Reading in Science 50

Chapter Vocabulary 52

Contents

Unit B Literature 54

Chapter 3 Interactions in Ecosystems
Chapter Concept Map 55
Lesson 1 Lesson Outline 56
   Lesson Vocabulary 58
   Lesson Cloze Activity 59
   Writing in Science 60
Lesson 2 Lesson Outline 62
   Lesson Vocabulary 64
   Lesson Cloze Activity 65
Lesson 3 Lesson Outline 66
   Lesson Vocabulary 68
   Lesson Cloze Activity 69
   Reading in Science 70
Chapter Vocabulary 72

Chapter 4 Ecosystems and Biomes
Chapter Concept Map 74
Lesson 1 Lesson Outline 75
   Lesson Vocabulary 77
   Lesson Cloze Activity 78
Lesson 2 Lesson Outline 79
   Lesson Vocabulary 81
   Lesson Cloze Activity 82
Lesson 3 Lesson Outline 83
   Lesson Vocabulary 85
   Lesson Cloze Activity 86
   Reading in Science 87
Lesson 4 Lesson Outline 89
   Lesson Vocabulary 91
   Lesson Cloze Activity 92
   Writing in Science 93
Chapter Vocabulary 95

Contents

EARTH SCIENCE

Unit C Literature 97

Chapter 5 Our Dynamic Earth
Chapter Concept Map.. 98
Lesson 1 Lesson Outline... 99
          Lesson Vocabulary 101
          Lesson Cloze Activity 102
Lesson 2 Lesson Outline...103
          Lesson Vocabulary 105
          Lesson Cloze Activity 106
          Writing in Science 107
Lesson 3 Lesson Outline...109
          Lesson Vocabulary 111
          Lesson Cloze Activity 112
Lesson 4 Lesson Outline... 113
          Lesson Vocabulary 115
          Lesson Cloze Activity 116
          Writing in Science 117
Lesson 5 Lesson Outline... 119
          Lesson Vocabulary 121
          Lesson Cloze Activity 122
          Reading in Science 123
Chapter Vocabulary 125

Chapter 6 Protecting Earth's Resources
Chapter Concept Map.. 127
Lesson 1 Lesson Outline...128
          Lesson Vocabulary 130
          Lesson Cloze Activity 131
Lesson 2 Lesson Outline...132
          Lesson Vocabulary 134
          Lesson Cloze Activity 135
Lesson 3 Lesson Outline...136
          Lesson Vocabulary 138
          Lesson Cloze Activity 139
          Writing in Science 140
Lesson 4 Lesson Outline...142
          Lesson Vocabulary 144
          Lesson Cloze Activity 145
          Reading in Science 146
Chapter Vocabulary 148

Contents

Unit D Literature 150

Chapter 7 Weather Patterns
Chapter Concept Map 151
Lesson 1 Lesson Outline 152
　　　　　　Lesson Vocabulary 154
　　　　　　Lesson Cloze Activity 155
Lesson 2 Lesson Outline 156
　　　　　　Lesson Vocabulary 158
　　　　　　Lesson Cloze Activity 159
Lesson 3 Lesson Outline 160
　　　　　　Lesson Vocabulary 162
　　　　　　Lesson Cloze Activity 163
　　　　　　Writing in Science 164
Lesson 4 Lesson Outline 166
　　　　　　Lesson Vocabulary 168
　　　　　　Lesson Cloze Activity 169
　　　　　　Reading in Science 170
Chapter Vocabulary 172

Chapter 8 The Universe
Chapter Concept Map 174
Lesson 1 Lesson Outline 175
　　　　　　Lesson Vocabulary 177
　　　　　　Lesson Cloze Activity 178
Lesson 2 Lesson Outline 179
　　　　　　Lesson Vocabulary 181
　　　　　　Lesson Cloze Activity 182
　　　　　　Writing in Science 183
Lesson 3 Lesson Outline 185
　　　　　　Lesson Vocabulary 187
　　　　　　Lesson Cloze Activity 188
　　　　　　Reading in Science 189
Lesson 4 Lesson Outline 191
　　　　　　Lesson Vocabulary 193
　　　　　　Lesson Cloze Activity 194
Chapter Vocabulary 195

Contents

PHYSICAL SCIENCE

Unit E Literature.. 197

Chapter 9 Comparing Kinds of Matter
Chapter Concept Map... 198
Lesson 1 Lesson Outline.. 199
         Lesson Vocabulary.. 201
         Lesson Cloze Activity..................................... 202
Lesson 2 Lesson Outline.. 203
         Lesson Vocabulary.. 205
         Lesson Cloze Activity..................................... 206
         Reading in Science....................................... 207
Lesson 3 Lesson Outline.. 209
         Lesson Vocabulary.. 211
         Lesson Cloze Activity..................................... 212
Chapter Vocabulary... 213

Chapter 10 Physical and Chemical Changes
Chapter Concept Map... 215
Lesson 1 Lesson Outline.. 216
         Lesson Vocabulary.. 218
         Lesson Cloze Activity..................................... 219
Lesson 2 Lesson Outline.. 220
         Lesson Vocabulary.. 222
         Lesson Cloze Activity..................................... 223
Lesson 3 Lesson Outline.. 224
         Lesson Vocabulary.. 226
         Lesson Cloze Activity..................................... 227
         Writing in Science.. 228
Lesson 4 Lesson Outline.. 230
         Lesson Vocabulary.. 232
         Lesson Cloze Activity..................................... 233
         Reading in Science....................................... 234
Chapter Vocabulary... 236

Contents

Unit F Literature	238
Chapter 11 Using Forces	
Chapter Concept Map	239
Lesson 1 Lesson Outline	240
Lesson Vocabulary	242
Lesson Cloze Activity	243
Reading in Science	244
Lesson 2 Lesson Outline	246
Lesson Vocabulary	248
Lesson Cloze Activity	249
Lesson 3 Lesson Outline	250
Lesson Vocabulary	252
Lesson Cloze Activity	253
Lesson 4 Lesson Outline	254
Lesson Vocabulary	256
Lesson Cloze Activity	257
Writing in Science	258
Chapter Vocabulary	260
Chapter 12 Using Energy	
Chapter Concept Map	262
Lesson 1 Lesson Outline	263
Lesson Vocabulary	265
Lesson Cloze Activity	266
Lesson 2 Lesson Outline	267
Lesson Vocabulary	269
Lesson Cloze Activity	270
Lesson 3 Lesson Outline	271
Lesson Vocabulary	273
Lesson Cloze Activity	274
Writing in Science	275
Lesson 4 Lesson Outline	277
Lesson Vocabulary	279
Lesson Cloze Activity	280
Reading in Science	281
Lesson 5 Lesson Outline	283
Lesson Vocabulary	285
Lesson Cloze Activity	286
Chapter Vocabulary	287

Name _____ Date _____

UNIT Literature

Adventures in Eating

Read the Literature feature in your textbook.

Write About It

Response to Literature This article tells about different adaptations for eating. Research two more animals that have interesting adaptations. Write a report that explains how these adaptations help the animals eat. Compare these adaptations to the ones you read about in the article.

_____
_____
_____
_____
_____
_____
_____
_____
_____
_____
_____
_____
_____
_____
_____
_____
_____
_____
_____

Unit A • Diversity of Life
Reading and Writing

1

CHAPTER Concept Map

Name _____ Date _____

Cells and Kingdoms

Complete the concept map by filling in answers where blanks appear.

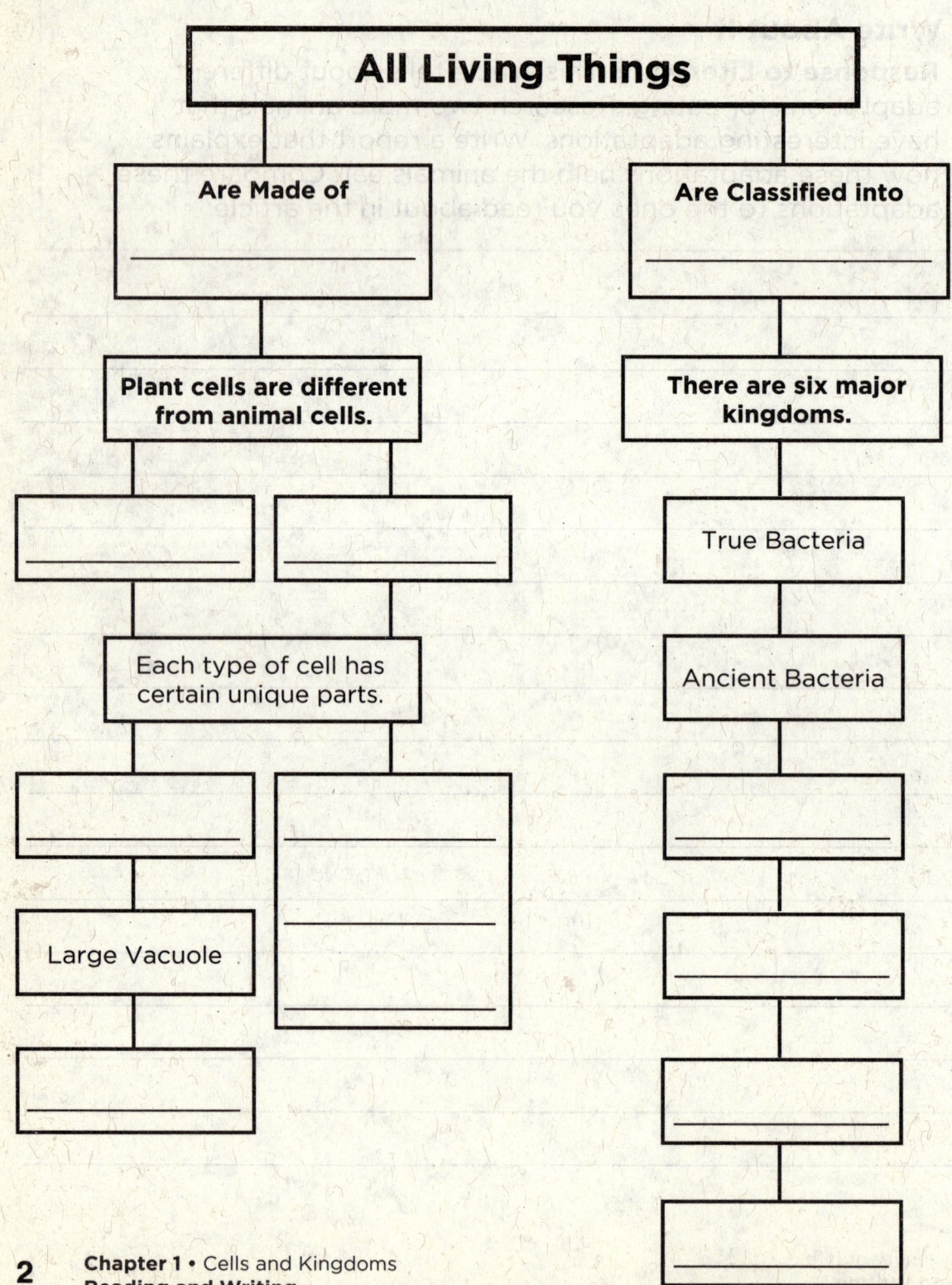

Chapter 1 • Cells and Kingdoms
Reading and Writing

Name _____ Date _____ **LESSON Outline**

Cells

Use your textbook to help you fill in the blanks.

What are cells?

1. All organisms, or living things, are made of _____ .

2. Every cell in every living thing comes from another cell that _____ .

3. A single-celled organism that can carry on all its life processes is called _____ .

4. Organisms made up of more than one cell are called _____ .

5. Scientists estimate that there are more than _____ kinds of unicellular organisms.

What is inside an animal cell?

6. Both plant and animal cells perform life processes by using _____ .

7. All cells are surrounded by a(n) _____ that controls the materials that move in and out of the cell.

8. The region between the cell membrane and the nucleus is filled with _____ .

9. The cell's control center is called the _____ .

Chapter 1 • Cells and Kingdoms
Reading and Writing

Use with **Lesson 1**
Cells

LESSON Outline

Name _____ Date _____

10. The tiny power plants in the cell where food is broken down and energy is released are called _____.

11. A structure in a cell used for storage of water, food, and waste is the _____.

What is inside a plant cell?

12. Plant cells have a(n) _____; a rigid structure that serves as an outer covering.

13. A green structure, called a(n) _____, uses the energy from the Sun to produce food for the plant.

How are cells organized?

14. Cells working together at the same job form a(n) _____.

15. Groups of tissues working together form organs, and groups of organs working together form _____.

Critical Thinking

16. Compare and contrast the cells of plants, animals, and unicellular organisms.

_____
_____
_____
_____
_____
_____
_____
_____

Name _____ Date _____

LESSON Vocabulary

Cells

Read each clue and fill in the crossword puzzle.

| cell | multicellular | organism | tissue |
| chlorophyll | organ | system | |

Across
2. The smallest unit of a living thing that can carry out the basic processes of life.
3. Similar cells working together at the same job.
5. A green chemical that absorbs sunlight.
7. Another name for a living thing.

Down
1. Organisms that contain many different types of cells.
4. Organs that work together to do a certain job.
6. A group of tissues working together to perform a specific job.

Chapter 1 • Cells and Kingdoms
Reading and Writing

Use with **Lesson 1**
Cells

5

LESSON Cloze Activity

Cells

Fill in the blanks.

cell membrane	cytoplasm	nucleus
cell wall	mitochondria	organelles
cells	multicellular	unicellular

When we talk about living things, we can break them down into smaller things. All living things are made up of units called _____ . Some organisms are _____ ; that is, they consist of only one cell. More complex organisms, including plants and animals, are called _____ organisms.

All cells are surrounded by a(n) _____ that controls what moves into and out of the cell. The insides of cells are filled with a gel like fluid called _____ . Within this liquid are the cell _____ . Both plant and animal cells, as well as many unicellular organisms, contain a(n) _____ and _____ , which supply energy for the cell. Plant cells have a(n) _____ , one large central vacuole, and chloroplasts. Chloroplasts contain chlorophyll which uses energy from sunlight to produce food for the plant.

Chapter 1 • Cells and Kingdoms
Reading and Writing

Use with **Lesson 1**
Cells

Name _____ Date _____

LESSON Outline

Classifying Life

Use your textbook to help you fill in the blanks.

How are organisms classified?

1. Scientists sort, or _____, living things into groups according to shared traits.

2. All organisms are divided into six major groups called _____.

3. The six subgroups used to classify organisms within kingdoms are _____, class, _____, family, genus, and _____.

4. The scientific name of an organism consists of its _____.

What are animals?

5. The two kingdoms that include only multicellular organisms are _____.

6. Plants can make their own food, but _____ obtain energy from other organisms.

7. The two major groups of animals are _____ and _____.

What are plants and fungi?

8. Although they are from two different kingdoms plants and fungi always have _____.

Chapter 1 • Cells and Kingdoms
Reading and Writing

Use with **Lesson 2**
Classifying Life

LESSON Outline   Name _____ Date _____

9. The two major groups into which plants are organized are _____ and _____.

10. Unlike plants, _____ get food by breaking down dead organisms.

11. A fungus that makes bread rise is called _____.

What are bacteria and protists?

12. Bacteria are unicellular organisms with no _____ or mitochondria.

13. The two kingdoms used to classify bacteria are the _____ and _____.

14. Protists can be _____ or multicellular.

15. Unlike bacteria, protists have large cells, a nucleus, and bound _____.

What are viruses?

16. Viruses are not classified as living organisms because they carry out no life processes except _____.

Critical Thinking

17. What makes plants and animals different from bacteria?

_____
_____
_____
_____

Name _____ Date _____

LESSON Vocabulary

Classifying Life

Match the correct letter with the description.

| **a.** classification | **c.** kingdom | **e.** species | **g.** vertebrate |
| **b.** invertebrate | **d.** nonvascular | **f.** vascular | **h.** virus |

1. _____ contains only closely related organisms

2. _____ the broadest group into which an organism can be classified

3. _____ helps scientists identify, study, group, and name organisms

4. _____ means "contains tubes or vessels"

5. _____ animal without a backbone

6. _____ type of plant that tends to be small and close to the ground

7. _____ animal with a backbone, nervous system, and brain

8. _____ organism that carries out only one life process, reproduction

Chapter 1 • Cells and Kingdoms
Reading and Writing

Use with **Lesson 2**
Classifying Life

LESSON Cloze Activity

Name _____ Date _____

Classifying Life

Fill in the blanks.

| classify | kingdom | species |
| genus | scientific name | unicellular |

Living things are often similar to each other. Scientists _____ living things by similarity into smaller and more specific groups. The largest group into which an organism is classified is its _____. The six kingdoms include plants, animals, fungi, protists, "true" bacteria, and "ancient" bacteria.

Each kingdom is divided into progressively smaller groups, as follows: phylum, class, order, family, genus, and species. The narrowest group into which an organism can be classified is a(n) _____. Two words make up the _____ of a living thing. The first is the _____ name. The second is the species name.

The "true" bacteria and the "ancient" bacteria are _____ organisms with no nucleus or mitochondria. "Ancient" bacteria usually live in very harsh environments in which other organisms could not survive.

Meet Angelique Corthals

Getting Ideas

Underline the sentence or sentences in each paragraph that state the main idea of the paragraph.

How can you fit thousands of organisms into one small room? Angelique Corthals knows. She's a scientist at the American Museum of Natural History, and she's been busy preserving tissue samples of many different organisms from around the globe, including samples from species that have become extinct or died out. Angelique is an archaeologist. That's a scientist who studies the past.

Angelique works in the museum's frozen tissues lab. She specializes in the preservation of the information in cells. From bacteria to insects to mammals, she collects, preserves, and organizes the cells of all sorts of living things. Angelique stores the organisms' cells and freezes them in small plastic tubes the size of your finger. Just as food stays fresh in the freezer, freezing cells prevents them from spoiling or decomposing. The tubes are stored in large tanks containing liquid nitrogen. At -150°C (-238°F), this liquid is so cold that all of the cells' biological processes stop.

By using this freezing process, the cells can be preserved for many years. When a scientist needs to study an organism, she can request a cell sample from the lab. Whether it's from a small fly or a large humpback whale, each cell contains information about the whole organism. Scientists can use this information to learn how different organisms are related. They can also use this information to learn about living things that have already become extinct and to understand why they died out.

Reading in Science

Name _____ Date _____

Write About It

Summarize Make a chart that tells the steps for preserving cells. Use your chart to write a summary of the process Angelique uses to freeze cells from organisms.

Sequence Make a chart that tells the steps for preserving cells. Use the blank boxes below.

First

↓

Next

↓

Last

Summarize In a paragraph, summarize the process that Corthals uses to freeze cells.

_____
_____
_____
_____

12 Chapter 1 • Cells and Kingdoms
Reading and Writing

Use with **Lesson 2**
Classifying Life

Name _____ Date _____

LESSON Outline

Plants

Use your textbook to help you fill in the blanks.

How are plants classified?

1. Small plants such as mosses which survive without a transport system, are called _____.

2. Plants that have a system of hollowed-out tubes to transport water and nutrients are called _____.

3. A seed plant that does not produce flowers or fruits is called a(n) _____.

4. A seed plant that produces flowers and some kind of fruit is called a(n) _____.

What are roots?

5. Roots absorb minerals and water, store food, and _____.

6. Root hairs absorb water and minerals, and _____ protect root tips.

7. The epidermis is on the outside of the root; just beneath it is the _____ which is used to store food. At the center of the root is the _____.

Chapter 1 • Cells and Kingdoms
Reading and Writing

Use with **Lesson 3**
Plants

LESSON Outline

Name _____ Date _____

What are stems?

8. Stems have two main functions: _____ and transport.

9. Grasses have _____ stems that are green and bendable; trees have _____ stems.

10. A series of tubes that move water and minerals up the plant are _____. _____ moves sugar made in the plant's leaves to other parts of the plant.

What are leaves?

11. The function of leaves is to perform _____.

12. To perform photosynthesis, chloroplasts need _____ from the air, water from the soil, and _____.

13. Air enters and exits plants through _____: pores on the underside of the leaves.

Critical Thinking

14. Why do you think some plants have woody stems and some have soft stems?

_____
_____
_____
_____
_____
_____

Name _____ Date _____

LESSON Vocabulary

Plants

Fill in the blank with a term from the box.

angiosperm	gymnosperm	transpiration
cambium	phloem	xylem
cellular respiration	photosynthesis	

1. A seed plant that does not produce flowers is called a(n) _____.

2. Cells that move sugars up, down, and all around a plant are called _____.

3. A layer in the plant stem that separates xylem and phloem is called the _____.

4. The break down of sugars in plant and animal cells to produce energy and carbon dioxide is called _____.

5. A seed plant that produces flowers is called a(n) _____.

6. Cells that transport water and minerals from roots to shoots in plants are called _____.

7. The process that plants use to produce their food and give off oxygen is called _____.

8. When water moves up the vascular tubes through stomata, _____ occurs.

Chapter 1 • Cells and Kingdoms
Reading and Writing

Use with Lesson 3
Plants

Plants

Fill in the blanks.

> angiosperms   gymnosperms   stomata
> cellular respiration   photosynthesis   sugar
> chloroplasts   stems

Seed plants can be divided into two main groups. Scientists call these groups _____ (flowering plants) and _____ (plants without flowers or fruits). Seed plants have three basic parts—leaves, roots, and _____ .

The function of leaves is to absorb sunlight to make sugars, a process called _____ . The energy of sunlight is captured by _____ and used to combine carbon dioxide and water. The carbon dioxide comes into the leaf through _____ . The sugar made during photosynthesis travels to cells all over the plant, where it is used for _____ . During this process, _____ is broken down to release energy to power the cell's functions. The by-products are carbon dioxide and water.

Name _____ Date _____

Writing in Science

Saving Water the Yucca Plant Way

Read the Writing in Science feature in your textbook.

Write About It

Explanatory Writing Write an article for young gardeners. Explain the process of CAM photosynthesis. Research facts and details for your article.

Planning and Organizing

Help Ray create an outline for his article. Here are some topics he wants to cover. Place them in the outline form below.

▶ What happens during the day in CAM photosynthesis?
▶ What is the purpose of CAM photosynthesis?
▶ What is photosynthesis?
▶ What happens at night during CAM photosynthesis?
▶ How does the process of CAM photosynthesis work?

I. _____

II. _____

III. _____

    A. _____

    B. _____

IV. Why is the yucca plant special?

Now create an outline for your own article on a separate sheet of paper. Make it as detailed as possible. Add A, B, C points and subpoints (1, 2, 3) under these as necessary.

Chapter 1 • Cells and Kingdoms
Reading and Writing

Use with **Lesson 3**
Plants

17

Writing in Science

Name _____ Date _____

Now use a separate sheet of paper to write the first draft of your article.

Revising and Proofreading

Here is part of the report that Ray wrote. Help him combine his sentences. Use the transition word in parentheses. Make sure you punctuate the new sentence correctly.

1. In CAM photosynthesis, the stomata open at night. The air is cooler and the humidity is higher. (when)

   _____
   _____

2. CAM photosynthesis is effective. It results in more efficient water use. (since)

   _____
   _____

Now revise and proofread your article. Ask yourself:

▶ Have I introduced my main idea about photosynthesis in yuccas?

▶ Have I included facts and details to show how this process works?

▶ Have I used examples and language appropriate for my audience?

▶ Have I used transition words and phrases to connect ideas?

▶ Have I ended with a strong conclusion about why yucca plants are special?

▶ Have I corrected all grammar errors?

▶ Have I corrected all problems in spelling, punctuation, and capitalization?

Chapter 1 • Cells and Kingdoms
Reading and Writing

Use with **Lesson 3**
Plants

Name _____ Date _____

LESSON Outline

Classifying Animals

Use your textbook to help you fill in the blanks.

What are simple invertebrates?

1. The simplest animals are _____.
   They are without real tissues or organs and have a(n)
   _____ body plan.

2. Jellyfish and hydras are _____.
   They possess a mouth and muscle tissue and are
   _____ symmetrical.

3. Worms that have flat bodies with one body opening
   and simple eyes are called _____.

4. Worms that have simple digestive and nervous systems
   are called _____.

What are complex invertebrates?

5. Clams and squids are _____. They have
   _____ symmetry, a muscular foot, a
   mantle, and several specialized organs.

6. Sea stars and sea cucumbers are _____.
   They have _____ feet and a water
   pressure system that helps them feed, breathe,
   and move.

7. Crabs and insects belong to the largest animal group on
   Earth, the _____ phylum.

Chapter 1 • Cells and Kingdoms
Reading and Writing

Use with Lesson 4
Classifying Animals

19

LESSON Outline

Name _____ Date _____

What are vertebrates?

8. There are three kinds of fish: _____, such as lamprey and hagfish; _____, such as sharks and skates; and _____.

9. Frogs, toads, and salamanders are _____.

10. Lizards, snakes, turtles, and alligators are _____. They are _____, which means that their body temperature is not steady.

11. Birds are designed for flying: they are warm-blooded and have _____ and feathers that are light and strong.

What are mammals?

12. Animals that are warm-blooded and have hair are called _____.

13. A duck-billed platypus lays eggs. It is a(n) _____.

14. A kangaroo is a(n) _____. It gives birth to partially developed offspring.

15. Lions, whales, and humans are _____. Their offspring develop within the mother.

Critical Thinking

16. Compare 4 different vertebrates.

_____
_____
_____
_____

Classifying Animals

Read each clue and fill in the blank with the correct answer.

| asymmetrical | invertebrates | monotreme | radial |
| bilateral | marsupial | placental | vertebrates |

1. _____ A koala is one. It gives birth to partially developed offspring.

2. _____ Worms have this kind of symmetry because they can be divided along only one plane.

3. _____ Fish, birds, amphibians, reptiles, and mammals.

4. _____ A whale is an example of this kind of mammal.

5. _____ Cnidarians have this kind of symmetry.

6. _____ A mammal that lays eggs.

7. _____ A type of body plan that has no definite shape.

8. _____ Sponges, cnidarians, echinoderms, mollusks, and arthropods.

LESSON Cloze Activity

Name _____ Date _____

Classifying Animals

Fill in the blanks.

| amphibians | invertebrates | sponges | vertebrates |
| hollow | reptiles | tentacles | |

The animal kingdom contains all the animals. The animal kingdom is separated into two large groups—animals with backbones called _____ and animals without backbones called _____ . These two groups are divided into smaller groups called phyla.

Vertebrates include fish, amphibians, reptiles, birds, and mammals. Fish live in the water and breathe through gills. Vertebrates that spend part of their lives in water and part on land are called _____ . Lizards, snakes, turtles, alligators, and crocodiles are _____ . Birds are designed for flying. Their bones are _____ and light. Mammals produce milk to feed their young.

Invertebrates include sponges, mollusks, worms, and arthropods. The most primitive of the animal groups are called _____ . Cnidarians have mouths surrounded by stinging _____ . The largest of all the animal groups are called arthropods.

Name _____ Date _____

Writing in Science

The Underground Life of Mole Rats

Write About It
Find out the scientific name of an animal you think is cute or ugly. Write a description of the animal. Use words and details that appeal to the senses in your description.

Getting Ideas

Choose an animal to describe. Then use the web below to brainstorm ideas. Write its scientific name in the center circle. Write details that describe it in the outer circles. You can add circles to the web if you like.

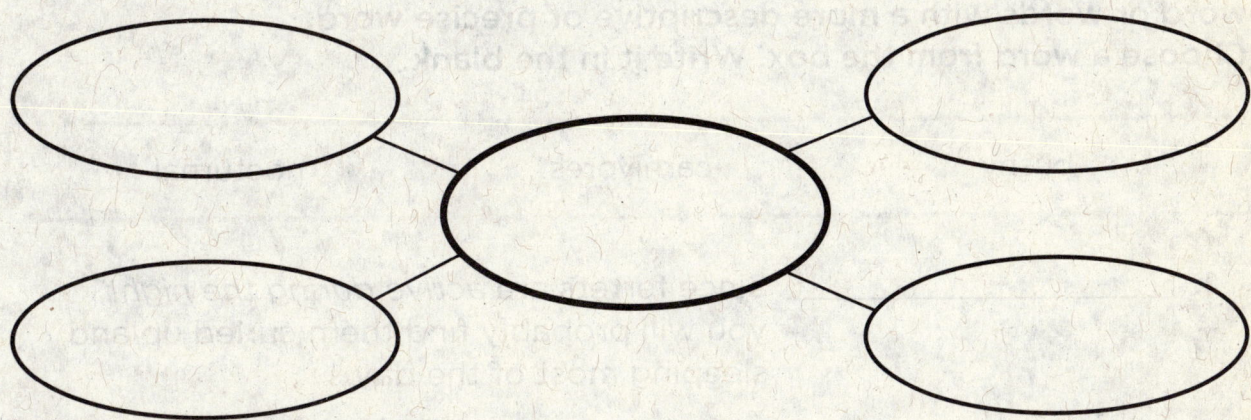

Planning and Organizing

Janine wants to describe a ferret. Here are some sentences that she wrote. Write Yes if the sentence contains words and details that create a vivid impression of the ferret. Write No if it does not.

1. _____ The dark black marks around its eyes make it look like a bandit.

2. _____ Its long, lean body curves upward as though in surprise.

3. _____ The ferret has four legs and a tail.

Chapter 1 • Cells and Kingdoms
Reading and Writing

Use with **Lesson 4**
Classifying Animals

Writing in Science

Name _____ Date _____

Drafting

Write a sentence to begin your description. Tell what animal you will describe. Make sure your sentence expresses the impression you want to create.

_____

_____

Now write your description. Use a separate piece of paper. Begin with the sentence you wrote above. Tell the animal's scientific name. Include descriptive words and vivid details to help readers experience the animal.

Revising and Proofreading

Here are three sentences that Janine wrote for her description. Help her improve them. Replace each italic word or words with a more descriptive or precise word. Choose a word from the box. Write it in the blank.

| bushy | carnivores | nocturnal |

1. _____ Since ferrets are *active during the night*, you will probably find them curled up and sleeping most of the day.

2. _____ Some ferrets have a *hairy* tail with an inky-black tip.

3. _____ Ferrets are *animals that eat meat*. They may eat squirrels, mice, and even prairie dogs.

Now revise and proofread your writing. Ask yourself:

▶ Did I create a vivid impression of the animal I chose?

▶ Did I use sensory words and vivid details?

▶ Did I correct all mistakes in grammar, spelling, punctuation, and capitalization?

Chapter 1 • Cells and Kingdoms
Reading and Writing

Use with **Lesson 4**
Classifying Animals

Name _____ Date _____

LESSON Outline

Animal Systems

Use your textbook to help you fill in the blanks.

What are the skeletal and muscular systems?

1. Bones, tendons, and ligaments make up the _____ system.

2. Muscles are attached to bones by _____ .

3. When a muscle receives a nerve signal, it _____ and pulls on a tendon.

What are the digestive and excretory systems?

4. From the mouth, food travels down the muscular _____ to the stomach.

5. Food is churned with strong acids in the _____ .

6. The system that removes wastes such as water, salt, carbon dioxide, and nitrogen from the body is called the _____ .

7. The blood carries wastes to the _____ , which filter the wastes from the blood.

What are the respiratory and circulatory systems?

8. The _____ and the passageways that lead to them make up the respiratory system in mammals.

9. The heart pumps oxygen-poor blood to the alveoli, where the oxygen enters the blood and _____ leaves the blood.

Chapter 1 • Cells and Kingdoms
Reading and Writing

Use with **Lesson 5**
Animal Systems

LESSON Outline

Name _____ Date _____

10. The blood travels to the small intestines and picks up _____ . Next, nutrient and oxygen-enriched blood travels through smaller and smaller blood vessels to body cells.

11. Nutrients and oxygen pass into the body cells and waste materials pass out of the cells into the blood through the _____ .

12. In vertebrates and some other animals, blood is carried in blood vessels. These animals have a(n) _____ . Arthropods and some other invertebrates have _____ circulatory systems.

What are the nervous and endocrine systems?

13. Your brain, spinal cord, nerves, and sense organs are part of your _____ .

14. The endocrine system sends out _____ that cause changes in your body.

Critical Thinking

15. Name two organ systems that work closely together and explain how they interact.

_____
_____
_____
_____
_____

Animal Systems

Who am I? What am I?

Match the correct letter with the description.

> **a.** circulatory system **d.** excretory system **g.** respiratory system
> **b.** digestive system **e.** muscular system **h.** skeletal system
> **c.** endocrine system **f.** nervous system

1. _____ I am a long tube in which food is broken down into nutrients your body can use. Who am I?

2. _____ I produce hormones that are released into your body to change body activity. Who am I?

3. _____ Through me, your blood takes in oxygen from the air.

4. _____ I cause your bones to move. Who am I?

5. _____ I carry oxygen and nutrients to all your body cells. Who am I?

6. _____ I remove wastes from your blood. Who am I?

7. _____ Because of me, you can see, hear, feel, taste, think, and move. Who am I?

8. _____ I am the framework for your body, and I protect your internal organs. Who am I?

LESSON Cloze Activity

Animal Systems

Fill in the blanks.

> circulatory system   endocrine system   nervous system
> digestive system   excretory system   respiratory system

Your body is made up of several organ systems that work together to carry out basic life functions. The system that is made up of the heart, blood, and blood vessels is called the _____. The system that brings air into the body and removes carbon dioxide is the _____.

A long tube in which food is broken down into the nutrients that the body needs for energy, growth, and repair is called the _____.

The _____ is made up of a group of glands that produce hormones. Hormones are chemicals that control body functions, such as heart rate. The brain, spinal cord, and nerves make up the _____, which sends, receives, and processes nerve impulses throughout the body.

The kidneys are a part of the _____. They eliminate waste from the blood and form urine.

Name _____ Date _____

CHAPTER Vocabulary

Cells and Kingdoms

Choose the letter of the best answer.

1. The flexible wrapping that surrounds all cells is called the
    a. cytoplasm.
    b. cell wall.
    c. cell membrane.
    d. vacuole.

2. Which of these structures is found in a plant cell, but not in an animal cell?
    a. central vacuole
    b. mitochondria
    c. nucleus
    d. organelles

3. Which of the following is in the correct order, from simplest to most complex?
    a. cells, tissues, organs, organism, organ systems
    b. cells, tissues, organs, organ systems, organism
    c. cells, organs, tissues, organ systems, organism
    d. cells, tissues, organ systems, organs, organism

4. Which phylum has the most species?
    a. echinodermata
    b. mollusca
    c. chordata
    d. arthropoda

5. In which kingdom do all of the members obtain energy by breaking down dead organisms?
    a. plants
    b. animals
    c. fungi
    d. protists

6. Which life process do viruses carry out?
    a. reproduction
    b. movement
    c. growth
    d. use of energy

7. Which structure found in vascular plants allows for the transport of water?
    a. flower
    b. xylem
    c. phloem
    d. seed

Chapter 1 • Cells and Kingdoms
Reading and Writing

CHAPTER Vocabulary Name _____ Date _____

Choose the letter of the best answer.

8. A fir tree is an example of a(n)
   a. nonvascular plant.
   b. angiosperm.
   c. seedless plant.
   d. gymnosperm.

9. A carrot root is an example of a(n)
   a. taproot.
   b. prop root.
   c. fibrous root.
   d. aerial root.

10. Which of the following moves sugars that are made in the plant's leaves?
    a. xylem
    b. phloem
    c. cambium
    d. bark

11. Which group includes the simplest animals?
    a. worms
    b. mollusks
    c. sponges
    d. cnidarians

12. Which animals have radial symmetry?
    a. sponges
    b. worms
    c. mollusks
    d. cnidarians

13. Which vertebrates have gills when they hatch, and lungs when they are adults?
    a. amphibians
    b. reptiles
    c. fish
    d. birds

14. Which organ filters wastes from the blood?
    a. bladder
    b. kidneys
    c. large intestine
    d. small intestine

15. Where does blood travel next after returning from the body cells?
    a. to the small intestine
    b. to the lungs
    c. to the liver
    d. to the kidneys

Name _____ Date _____

CHAPTER Concept Map

Parents and Offspring

Complete the concept map with information you have learned about different types of reproduction. Some answers have been written for you.

All Living Things Reproduce

Types of Reproduction	Organisms that use this type of reproduction	Does this type of reproduction enhance genetic variation?	Disadvantages or Advantages to this type of reproduction
_____			_____ _____ _____
_____	bacteria	_____	_____ _____ _____

Chapter 2 • Parents and Offspring
Reading and Writing

Reproduction

Use your textbook to help you fill in the blanks.

What are sexual and asexual reproduction?

1. Survival of a(n) _____ depends on its ability to produce offspring.

2. Every organism comes from a parent through the process of _____ .

3. The transfer of _____ from parents to their offspring is known as reproduction.

4. Genetic material contains the information that controls an organism's _____ .

5. The production of a new organism from two parents is called _____ reproduction.

6. When an egg cell joins with a sperm cell, _____ occurs.

7. A fertilized egg develops into an individual with traits from each _____ .

8. The production of a new organism from a single parent is called _____ reproduction.

How do organisms reproduce asexually?

9. Most bacteria and unicellular protists reproduce by making a copy of their genetic material and _____ .

10. Cnidarians, sponges, and some fungi can reproduce through _____ .

Name _____ Date _____

LESSON Outline

11. The eggs of insects, fish, frogs, and lizards sometimes develop into new animals without being _____.

12. New plants can grow from leaves, roots, or stems. This type of asexual reproduction is called _____.

13. Strawberry plants and ferns can reproduce asexually by forming _____.

How do sexual and asexual reproduction compare?

14. An organism that reproduces asexually does not have to find a(n) _____.

15. Organisms that reproduce asexually tend to be well suited to their _____.

16. A major advantage of sexual reproduction is that it promotes _____ in a species.

Critical Thinking

17. Why is sexual reproduction better than asexual reproduction for ensuring the survival of a species in a changing environment?

_____
_____
_____
_____
_____
_____

Chapter 2 • Parents and Offspring
Reading and Writing

Use with **Lesson 1**
Reproduction

LESSON Vocabulary

Reproduction

Read each clue. Write the answer in the blanks using the words below. Then fill in the crossword puzzle.

> asexual   runners   splitting   variation
> budding   sexual   trait   vegetative

Across

3. Plant stems that run along the ground and sprout as new plants.
5. Any characteristic of a living thing.
6. The production of a new organism from one parent is called _____ reproduction.
7. Bacteria reproduce by _____.

8. Sexual reproduction gives rise to _____ in a species.

Down

1. A bud growing from a fungus to become a new individual.
2. A new plant growing from a leaf is _____ propagation.
4. A new organism from two parents is _____ reproduction.

Chapter 2 • Parents and Offspring
Reading and Writing

Use with Lesson 1
Reproduction

Reproduction

Fill in the blanks.

| asexual | reproduce | sperm | variety |
| mate | sexual | splitting | |

No organism lives forever. This means all organisms must _____ . There are two types of reproduction: _____ and _____ . Sexual reproduction requires two parents. A female egg cell unites with a male _____ cell to produce a fertilized egg. The fertilized egg grows into a new, unique individual. Asexual reproduction requires only one parent and results in offspring that are genetically identical to the parent.

The main advantage of sexual reproduction is that it promotes _____ within a species. An advantage of asexual reproduction is that it does not require finding a(n) _____ . There are several methods of asexual reproduction. Simple, one-celled organisms, such as bacteria and protists, reproduce by _____ into two cells. Animals such as cnidarians and sponges undergo a process called budding.

Writing in Science

Name _____ Date _____

How Do Sea Stars Regenerate?

Write About It

The article you just read explained how the sea star can produce offspring using regeneration. Choose another animal that can reproduce without two parents. Write an explanation of how this process takes place.

Getting Ideas

Choose an animal to write about. Think about how it reproduces without parents. Write the steps below.

First

↓

Next

↓

Last

Planning and Organizing

Xavier wants to explain how flat worms reproduce. Here are three sentences he wrote. Put them in order.

_____ Finally, each half grows into a separate flat worm.

_____ First, the flat worm divides in two.

_____ Stem cells turn into the types of cells needed to reproduce the missing part.

Name _____ Date _____

Writing in Science

Drafting

Write a sentence to begin your explanation. Name the animal you are writing about. Tell your main idea about how this animal reproduces. This is your topic sentence.

_____

_____

_____

Now write your explanation. Use a separate piece of paper. Begin with your topic sentence. Explain how the animal reproduces. Write the steps in time order.

Revising and Proofreading

Here are some sentences Xavier wrote. Combine each pair. Use the time order word in parentheses. Write the new sentence on the line.

1. The stem cells multiple. They turn into specialized cells. (before)

   _____

   _____

2. A message is sent out to specialized cells. The cells near the wound cover it. (after)

   _____

   _____

Now revise and proofread your writing. Ask yourself:

▶ Did I explain how the animal can reproduce without parents?

▶ Did I include time order words?

▶ Did I correct all mistakes?

Chapter 2 • Parents and Offspring
Reading and Writing

Use with **Lesson 1** Reproduction

LESSON Outline Name _____ Date _____

Plant Life Cycles

Use your textbook to help you fill in the blanks.

What are seedless plant life cycles?

1. Plant life cycles have two alternating phases, one sexual and one asexual. This type of life cycle is called _____.

2. During the asexual phase, moss plants form capsules that contain _____.

3. During the sexual phase, moss spores grow into plants with male and female branches. Rainwater carries sperm to egg cells, and _____ occurs.

What are the parts of a flower?

4. The male part of a flower is called the _____; the female part is called a pistil.

5. At the top of the filament is the _____, where pollen is produced.

6. The pistil is made up of a stigma, a style, and a(n) _____ (which contains the egg cells).

7. A perfect flower has both a stamen and a pistil; a(n) _____ flower lacks one part or the other.

What is an angiosperm life cycle?

8. The transfer of pollen from stamen to pistil is called _____.

Name _____ Date _____

LESSON Outline

9. After pollination, sperm cells from pollen move down the _____ of the pistil to the ovary.

What is in a seed?

10. The ovary enlarges to become a(n) _____ as the seeds develop.

11. In addition to the embryo, a seed contains a food supply called the _____ .

12. The development of a seed into a new plant is called _____ .

What is the conifer life cycle?

13. Cone-bearing plants, such as pines and firs, are called _____ .

14. After a conifer egg is fertilized, it develops into a seed on the _____ .

Critical Thinking

15. Compare and contrast the reproduction of mosses, ferns, gymnosperms, and angiosperms.

_____
_____
_____
_____
_____
_____
_____

Chapter 2 • Parents and Offspring
Reading and Writing

Use with **Lesson 2**
Plant Life Cycles

LESSON Vocabulary

Plant Life Cycles

Who am I? What am I?

Choose a word from the word box that answers each question.

a. conifer	c. embryo	e. monocot	g. pollination
b. dicot	d. germination	f. pollen	h. seed coat

1. _____ I am a cone-bearing tree. Who am I?

2. _____ I take place when pollen from the stamen reaches the pistil. What am I?

3. _____ I am the tiny offspring inside a seed that can grow into a new plant. Who am I?

4. _____ I am the development of a seed into a new plant. What am I?

5. _____ I contain a flowering plant's sperm cells. Who am I?

6. _____ I have seeds with two cotyledons. Who am I?

7. _____ I have seeds with one cotyledon. Who am I?

8. _____ I am the tough, outer covering on a seed. What am I?

Plant Life Cycles

Fill in the blanks.

> alternation of generations seeds
> cones sexual phase
> flowers spore cases
> pollination spores

All plants have a life cycle with two phases—one sexual and one asexual. This type of life cycle is called _____ . In gymnosperms and angiosperms, the asexual phase is much reduced, and the _____ is the dominant part of the life cycle. Gymnosperms produce male and female _____ . When pollen from the male cones reaches the female cones, _____ occurs. The fertilized eggs stay attached to the female cones as they develop into _____ . Angiosperms produce reproductive organs called _____ .

Moss and fern plants produce _____ during their asexual phases. In ferns, spores are produced in _____ on the underside of the fronds. When the eggs are fertilized, they grow into fern fronds.

LESSON Outline

Name _____ Date _____

Animal Life Cycles

Use your textbook to help you fill in the blanks.

What are animal life cycles?

1. Some animals go through a series of distinct growth stages called _____ .

2. A butterfly hatches from an egg as a larva. The larva feeds and grows until it forms a(n) _____ .

3. Inside the pupa, a larva's body changes completely into a(n) _____ butterfly.

4. Grasshoppers emerge from their eggs as _____ , which are similar to the adult but lack wings and reproductive organs.

How does fertilization occur in animals?

5. Sperm and egg cells must stay protected and _____ for fertilization to occur.

6. Fish and amphibians release their sex cells into the surrounding water, where _____ fertilization takes place.

7. Fish and amphibians must release large amounts of sex cells because the chances of _____ in open water are low.

8. The joining of sperm and egg cells inside the female's body is called _____ .

42 Chapter 2 • Parents and Offspring
Reading and Writing

Use with Lesson 3
Animal Life Cycles

Name _____ Date _____

LESSON Outline

9. Animals that use internal fertilization include
   _____ , birds, and mammals.

10. Internal fertilization increases the chances that eggs will be
    _____ and offspring will survive.

What happens to a fertilized egg?

11. Animals that lay their eggs in open water include fish
    and _____ .

12. The egg's _____ provides food for a developing embryo.

13. Reptiles and birds have eggs filled with a liquid and
    surrounded by a tough _____ , so their eggs
    can be laid on land.

14. The embryos of most _____ develop inside
    the mother.

Critical Thinking

15. Compare and contrast complete and incomplete metamorphosis. Give an example of an organism that undergoes each.

    _____
    _____
    _____
    _____
    _____
    _____
    _____

Chapter 2 • Parents and Offspring
Reading and Writing

Use with **Lesson 3**
Animal Life Cycles

LESSON Vocabulary

Name _____ Date _____

Animal Life Cycles

Read each clue. Write the answer in the blanks using the words below. Then fill in the crossword puzzle.

| complete | incomplete | larva | nymph |
| external | internal | metamorphosis | pupa |

Across

1. The immature stage that emerges from the egg during incomplete metamorphosis.
4. Larva changes to an an adult inside this hard case.
5. A life cycle with three growth stages.
7. The immature stage that emerges from the egg during complete metamorphosis.

Down

2. A series of distinct growth stages.
3. A life cycle with four very distinct growth stages is called _____ metamorphosis.
5. The joining of egg and sperm cells inside the body.
6. The joining of egg and sperm cells outside the body.

Chapter 2 • Parents and Offspring
Reading and Writing

Use with **Lesson 3**
Animal Life Cycles

Name _____ Date _____

LESSON Cloze Activity

Animal Life Cycles

Fill in the blanks.

complete	internal	sperm
embryos	jelly-like layer	tough shells
external	metamorphosis	yolk

Animals reproduce sexually. Sexual reproduction of animals begins when egg and _____ cells unite. Fish and amphibian eggs are fertilized outside the female's body, a process called _____ fertilization. Land animals rely on _____ fertilization.

After egg cells are fertilized, they develop into _____. Fish and frog embryos develop inside soft eggs. The eggs are somewhat protected by a(n) _____ around them. Reptiles and birds lay eggs covered by _____. Their eggs contain an embryo, a watery fluid, and a food source, the _____.

When most animals are born, they look like their parents. Other animals go through a series of stages called _____. Butterflies, moths, and beetles go through _____ metamorphosis. Grasshoppers, and termites go through incomplete metamorphosis.

Chapter 2 • Parents and Offspring
Reading and Writing

Use with **Lesson 3**
Animal Life Cycles

LESSON Outline   Name _____   Date _____

Traits and Heredity

Use your textbook to help you fill in the blanks.

What is heredity?

1. The passing of traits from one generation to the next is called _____ .

2. Traits that offspring receive from their parents are _____ traits.

3. A way of acting or behaving with which an animal is born is called a(n) _____ .

4. A behavior that develops during an animal's lifetime is a(n) _____ behavior.

5. When ducks hatch, they learn to recognize and follow their mother, a behavior called _____ .

How are traits inherited?

6. Mendel discovered that each inherited trait is controlled by _____ , one from each parent.

7. Today scientists refer to Mendel's factors as _____ .

8. Genes are found in the nucleus of the cell. They are stored on _____ .

9. A trait that masks another trait is called a(n) _____ trait.

10. A trait that is masked is called a _____ trait.

46 Chapter 2 • Parents and Offspring
Reading and Writing

Use with **Lesson 4**
Traits and Heredity

Name _____ Date _____

LESSON Outline

11. In pea plants, purple flowers are a dominant trait and white flowers are a recessive trait. The purple trait is represented by _____ and the white trait by p.

How do we trace inherited genes?

12. A chart used to trace the history of traits in a family is called a(n) _____ .

13. On a pedigree chart, horizontal lines connect parents and vertical lines connect parents to _____ .

14. Males are represented by squares, and _____ are represented by circles.

15. Shaded shapes represent individuals with a particular _____ , and unshaded shapes represent individuals without that trait.

16. Dimples are a dominant trait, represented by the letter D. A child who is a carrier of the recessive trait is represented by _____ .

Critical Thinking

17. Both a father and mother have dimples. Their son has dimples, but their daughter does not. Which genes, DD, Dd, or dd, does each family member have?

_____
_____
_____
_____
_____

Chapter 2 • Parents and Offspring
Reading and Writing

Use with **Lesson 4**
Traits and Heredity

47

LESSON Vocabulary

Traits and Heredity

Match the correct letter with the description.

a. carrier	d. heredity	g. pedigree
b. dominant	e. inherited	h. recessive
c. gene	f. instinct	

1. _____ a trait that an offspring receives from its parents

2. _____ the passing down of traits from one generation to the next

3. _____ behavior that is inherited

4. _____ a trait that masks another trait

5. _____ a trait that is masked or covered by another trait

6. _____ chart used to trace the history of traits in a family

7. _____ contains the chemical instructions for an inherited trait

8. _____ individual who has inherited a gene for a trait, but does not show the trait physically

LESSON Cloze Activity

Name _____ Date _____

Traits and Heredity

Fill in the blanks.

| chromosomes | heredity | Gregor Mendel | sperm cell |
| genes | instincts | pedigree | trait |

Parents pass on features of themselves to their offspring. Any notable feature of an organism is called a(n) _____ . The passing down of traits from parents to offspring is called _____ . Some traits, such as hair or eye color, are physical traits. Other inherited traits are behavioral and are called _____ . An Austrian monk, _____ , discovered how traits are inherited.

Today, Mendel's factors are called _____ . They are stored on the _____ inside the nucleus of cells. Offspring receive one set of genes from an egg cell and the other from the _____ that fertilized the egg cell.

Humans have an estimated 20,000 gene pairs. Some of these traits are easy to see. The history of a family trait and the way it has been inherited can be charted in a _____ . These charts can be used to study heredity patterns.

Chapter 2 • Parents and Offspring
Reading and Writing

Use with Lesson 4
Traits and Heredity

49

Reading in Science

Name _____ Date _____

Genetically Modified Corn

Read the Reading in Science feature from your textbook. Look for cause and effect relationships.

Cause and Effect

Fill in the Cause and Effect Chart with cause and effect relationships you find in the article.

Cause	Effect
Corn borer eats corn	
Bt powder sprayed on corn	
	Corn plants make Bt toxin in their own cells, so the corn plants can protect themselves.
Other living things eat Bt corn	

50 Chapter 2 • Parents and Offspring
Reading and Writing

Use with **Lesson 4**
Traits and Heredity

Name _____ Date _____ **Reading in Science**

 **Write About It**

Cause and Effect Explain how the bacterium Bt affects corn borers. Tell how genetically modified corn might cause problems for other insects and for the environment in general.

Planning and Organizing
Answer these questions in detail.

1. What does the Bt bacterium produce, and what effect does it have on corn borers?

   _____
   _____

2. What enables the Bt bacterium to make a protein that is toxic to corn borers?

   _____
   _____

3. What was transferred from the Bt bacterium to Bt corn?

   _____
   _____

4. How does Bt corn affect corn borers?

   _____
   _____

5. How might Bt corn affect other living things, such as monarch butterflies?

   _____
   _____

Chapter 2 • Parents and Offspring
Reading and Writing

Use with **Lesson 4**
Traits and Heredity

CHAPTER Vocabulary

Name _____ Date _____

Parents and Offspring

Choose the letter of the best answer.

1. Which of the following organisms reproduces by using budding?

   a. sponge   c. lizard
   b. cat   d. frog

2. Which of the following plants reproduces by using runners?

   a. corn plant
   b. moss
   c. strawberry plant
   d. apple tree

3. Which of the following is an example of sexual reproduction?

   a. cloning
   b. budding
   c. seed production
   d. vegetative propagation

4. Which organisms can develop from an unfertilized egg?

   a. humans   c. some birds
   b. all sheep   d. certain lizards

5. Which of the following is an advantage of asexual reproduction?

   a. It depends on finding another organism.
   b. It promotes variety in a species.
   c. It is convenient.
   d. It gives rise to offspring better suited to environmental change.

6. Where on a flower is pollen made?

   a. stigma   c. anther
   b. style   d. pistil

7. Where on a plant are egg cells produced?

   a. ovary   c. anther
   b. pistil   d. filament

8. When a new plant sprouts from a seed, it is

   a. fertilizing.
   b. pollinating.
   c. beginning its asexual phase.
   d. germinating.

Name _____ Date _____

**CHAPTER
Vocabulary**

Choose the letter of the best answer.

9. A flower with small, dull petals is most likely pollinated by
   a. birds.
   b. wind.
   c. bats.
   d. insects.

10. A dandelion seed is dispersed by
    a. clinging to the fur of animals.
    b. water.
    c. wind.
    d. being eaten by animals.

11. What is one of the main differences between a gymnosperm and an angiosperm?
    a. Only angiosperms produce seeds.
    b. Only angiosperms have leaves.
    c. Only angiosperms produce pollen.
    d. Only angiosperms produce fruits.

12. Which insect undergoes complete metamorphosis?
    a. beetle
    b. dragonfly
    c. bed bug
    d. grasshopper

13. Which of the following animals uses external fertilization?
    a. bird
    b. frog
    c. bear
    d. butterfly

14. Which of the following insects is a nymph at some point in its life cycle?
    a. moth
    b. grasshopper
    c. fly
    d. beetle

15. Which of the following items represents a carrier for the recessive trait?
    a. DD
    b. Dd
    c. dd
    d. d

16. An instinct is an example of
    a. a learned behavior.
    b. an inherited behavior.
    c. an inherited physical trait.
    d. imprinting.

17. If purple is the dominant gene for flower color, which of following items represents a white flower?
    a. PP
    b. pp
    c. Pp
    d. p

**Chapter 2 • Parents and Offspring
Reading and Writing**

53

UNIT **Literature**

Name _____ Date _____

The Case for Clean Water

Read the Literature feature in your textbook.

Write About It

Response to Literature This article tells how to find out if a body of water is clean. Research additional information about the insect larvae mentioned in the article. Write a report about the effects of pollution on these insects. Include facts and details from the article and from your research.

_____
_____
_____
_____
_____
_____
_____
_____
_____
_____
_____
_____
_____
_____
_____
_____
_____
_____

Unit B • Interactions in Ecosystems
Reading and Writing

Name _____ Date _____

CHAPTER Concept Map

Interactions in Ecosystems

Complete the concept map about relationships within ecosystems.

```
        ┌─────────────────────┐
        │ _____ Factors   │
        └─────────────────────┘
                  │
                such as
                  ↓
    ┌──────────────────────────────┐
    │ • _____     • Space     │
    │ • Soil                       │
    │                  • _____│
    │ • _____     • Temperature│
    └──────────────────────────────┘
                  │
              determine the
                  ↓
        ┌─────────────────────┐
        │ _____ Capacity  │
        └─────────────────────┘
                  │
               of a given
                  ↓
            ┌───────────┐
            │ Ecosystem │
            └───────────┘
                  │
            which is made up of
                  ↓
      ┌────────────────────────┐
      │ _____ of Organisms │
      └────────────────────────┘
                  │
            in which are found
                  ↓
        ┌─────────────────────┐
        │ • Producers         │
        │                     │
        │ • _____        │
        │                     │
        │ • _____        │
        └─────────────────────┘
```

Chapter 3 • Interactions in Ecosystems
Reading and Writing

55

LESSON Outline

Name _____ Date _____

Energy Flow in Ecosystems

Use your textbook to help you fill in the blanks.

What is an ecosystem?

1. The living things in an environment are _____ factors.

2. The nonliving things in an environment are _____ factors.

3. All the living and nonliving things interacting in an environment make up a(n) _____ .

4. All the members of a species within an ecosystem are a(n) _____ .

5. Together, the populations in an ecosystem form a(n) _____ .

How are food chains alike?

6. The path that energy takes in an ecosystem as it flows from organism to organism is a(n) _____ .

7. At the base of each food chain are _____ that use the Sun's energy to make sugar and oxygen during _____ .

8. The sugars provide food for _____ , or plant-eating animals.

Name _____ Date _____

LESSON Outline

9. Organisms in an ecosystem that break down dead or decaying plants and animals are _____.

10. Animals such as vultures and raccoons are _____ that eat dead bodies after they have started to rot.

What are food webs made of?

11. A network of food chains that share some links is a(n) _____.

How do energy pyramids compare?

12. A diagram that shows the energy that is available at each level of an ecosystem is a(n) _____.

13. At each level of an energy pyramid, about _____ percent of the energy from the level below is lost.

How does change affect a food web?

14. Removing a species from a food web can throw an ecosystem out of _____.

Critical Thinking

15. What would happen if producers were removed from an ecosystem?

_____
_____
_____
_____
_____

Chapter 3 • Interactions in Ecosystems
Reading and Writing

Use with **Lesson 1**
Energy Flow in Ecosystems

LESSON Vocabulary

Name _____ Date _____

Energy Flow in Ecosystems

Who am I? What am I?

Choose a word from the word box that answers each question.

a. community	d. food chain	g. predator
b. ecosystem	e. food web	h. prey
c. energy pyramid	f. population	

1. _____ I include all living things in an ecosystem. What am I?

2. _____ I am a diagram that shows the amount of energy available at each level of an ecosystem. What am I?

3. _____ I am a network of food chains that are connected. What am I?

4. _____ I am an animal that hunts other animals for food. Who am I?

5. _____ I include all living and nonliving things in an environment. What am I?

6. _____ Predators hunt me for food. Who am I?

7. _____ All the members of a single species in an ecosystem are part of me. What am I?

8. _____ I am the path that energy takes as it moves from one organism to another in an ecosystem. What am I?

58 Chapter 3 • Interactions in Ecosystems
Reading and Writing

Use with **Lesson 1**
Energy Flow in Ecosystems

Energy Flow in Ecosystems

Fill in the blanks.

| carnivores | food chain | herbivores | plants |
| community | food web | omnivores | population |

All the living and nonliving things in an environment make up an ecosystem. Within an ecosystem, all living things make up a _____. All individuals of one species are a(n) _____. An ecosystem can be as large as a forest or as small as a fallen log.

The path that energy takes as it moves from one organism to another in an ecosystem is a(n) _____. A group of connected food chains is a(n) _____. Producers, such as _____ and algae, are at the base of each food chain. Consumers include _____ that eat plants and _____ that eat other animals. Animals that eat both plants and animals are _____. The amount of energy available at each level in an ecosystem is shown by an energy pyramid.

Writing in Science

Name _____ Date _____

Two Desert Creatures

Write About It
Choose two other organisms that share a predator/prey relationship. Write a fictional narrative in which these two organisms are in a conflict.

Getting Ideas
Select two other animals. Think about these questions:
What is the problem? What happens between them?
Then use the sequence chart below to plan your story.

First

↓

Next

↓

Last

Planning and Organizing
Andy wanted to write about a red-tailed hawk and a muskrat. Here are three sentences that he wrote. Write 1 by the event that happens first. Write 2 by the event that happens second. Write 3 by the event that happens last.

_____ The hawk swooped down, grabbed the muskrat with its sharp talons, and carried it away.

_____ The hawk spied a large muskrat coming out of its burrow by the bank of the river.

_____ Seeing the hawk, the muskrat jumped in the water and tried to paddle away.

Name _____ Date _____

Writing in Science

Drafting

Write a sentence to begin your fictional narrative. Introduce the predator. Tell where the story takes place. Tell what the problem is.

_____

_____

Now write your fictional narrative. Use a separate piece of paper. Begin with the sentence you wrote above. Explain the conflict, or problem, between the predator and prey, and show how it is resolved. Tell these events in time order. Include dialogue to bring your characters to life.

Revising and Proofreading

Here are two sentences that Andy wrote. Each sentence is missing two punctuation marks. Rewrite them, adding punctuation marks where needed.

1. "Don't be afraid little muskrat, shouted the hawk, I just want to be your friend."

   _____

   _____

2. As the hawks powerful wings and large body threw a shadow over the land the muskrat looked up in fear.

   _____

   _____

Now revise and proofread your writing. Ask yourself:

▶ Did I include details that bring my characters to life?

▶ Did I present a reasonable conflict and show how it was resolved?

▶ Did I correct all mistakes in grammar, spelling, capitalization, and punctuation?

Chapter 3 • Interactions in Ecosystems
Reading and Writing

Use with **Lesson 1**
Energy Flow in Ecosystems

Relationships in Ecosystems

Use your textbook to help you fill in the blanks.

Why do organisms compete?

1. The struggle for resources among organisms in an ecosystem is called _____.

2. Any resource that restricts the growth of populations in an ecosystem is a(n) _____.

3. The size of the population that an area can support is its _____.

How do organisms avoid competition?

4. An organism's _____ is the place in which it lives and hunts for food.

5. The specific role that an organism plays within a community is that organism's _____.

How do organisms benefit from interactions?

6. The reliance of organisms on one another for survival is known as _____.

7. A relationship between organisms that lasts over time is _____.

8. A symbiotic relationship in which both organisms benefit is _____.

Name _____ Date _____

LESSON Outline

9. One example of this relationship is the _____, which is formed by a(n) _____ and an alga that live together.

10. A symbiotic relationship in which one organism benefits and the other is not harmed is _____.

11. One example of this type of relationship is the growth of _____ on the backs of whales; in this situation, no harm comes to the whales.

What are parasites?

12. A symbiotic relationship in which one organism benefits while the other is harmed is _____.

13. In this type of relationship, a(n) _____ benefits by living in or on a(n) _____.

14. Some parasites cause serious problems, giving people _____ such as dysentery.

Critical Thinking

15. What keeps populations in a community from increasing constantly?

_____
_____
_____
_____
_____

LESSON Vocabulary

Name _____ Date _____

Relationships in Ecosystems

Fill in the blanks.

a. carrying capacity	d. limiting factor	g. parasitism
b. commensalism	e. mutualism	h. symbiosis
c. habitat	f. niche	

1. The particular role that an organism plays in a community is its _____ .

2. When two organisms benefit in a symbiotic relationship, the relationship is called _____ .

3. The place in which an organism lives and hunts for food is its _____ .

4. Water is a(n) _____ that restricts the growth of populations in an ecosystem.

5. A relationship in which one organism benefits and the other is not harmed is _____ .

6. Because each area has a certain _____ , it can support only a limited population.

7. A symbiotic relationship in which one organism benefits while the other is harmed is _____ .

8. A special relationship between organisms that lasts a long time is _____ .

Relationships in Ecosystems

Fill in the blanks.

```
carrying capacity          host
commensalism               parasitism
compete                    symbiosis
exceeds                    vegetation
```

Each ecosystem has certain limiting factors that restrict the size of its populations. These include water, temperature, soil types, and the amount of _____ available for food. The population that any area can support is its _____. When the population of an area _____ its carrying capacity, some plants or animals begin to die off.

Living things _____ for resources in an ecosystem. However, _____ limits competition as organisms develop relationships that allow them to live together. A symbiotic relationship that benefits only one organism but does no harm to the other is known as _____. In _____, a parasite harms the _____ organism it lives on or in. In the relationship called mutualism, both organisms benefit.

Adaptation and Survival

Use your textbook to help you fill in the blanks.

What is adaptation?

1. A characteristic that helps an organism survive in its natural environment is a(n) _____ .

2. Organisms that are best adapted to their environment _____ and pass on their traits to offspring.

3. A physical structure that helps an organism survive in its environment, such as the _____ of an animal's fur, is a(n) _____ adaptation.

4. A characteristic that is an organism's response to its environment is a(n) _____ adaptation.

What are some plant adaptations?

5. Some plants have adapted to _____ environments by developing thick, waxy stems to prevent water loss.

6. Plants that are common in cold climates often have _____ growing periods.

7. Some plants produce bad-tasting chemicals that make them unattractive to _____ that might eat them.

Name _____ Date _____ **LESSON Outline**

What are some animal adaptations?

8. To keep warm in cold climates, animals have _____ fur.

9. In hot deserts, animals are often more active at _____ , when temperatures drop.

10. Any color, shape, or pattern that lets an organism blend into its environment is _____ .

11. A type of camouflage in which an organism's coloring helps it blend in with its background is _____ coloring.

12. When an organism matches the color, shape, and texture of the environment around it, it is showing protective _____ .

What is mimicry?

13. An adaptation in which an organism gets protection from predators by looking like a dangerous animal is _____ .

14. Predators also use this characteristic to fool _____ ; believing that the predators are harmless, prey come close enough to be caught.

Critical Thinking

15. How do adaptations help an organism survive in its environment?

_____
_____
_____

Chapter 3 • Interactions in Ecosystems
Reading and Writing

Use with **Lesson 3**
Adaptation and Survival

LESSON Vocabulary

Name _____ Date _____

Adaptation and Survival

Use the clues below to help you find the words hidden in the puzzle.

```
R R E S E M B L A N C E
Z M R T I G A D B Y A G
E K Y T H J L L X X M J
C Z Z S S L L M R T O M
C O L O R A T I O N U D
H I Q K P W W M B P F H
A D A P T A T I O N L N
P S V M M R G C A A A N
X F F L I U U R D T G K
Z H R T P I U Y I R E O
```

1. An organism that matches the color, shape, and texture of its environment is using protective _____.

2. A type of coloring, shape, or pattern that allows an organism to blend in with its environment is _____.

3. Any characteristic that helps an organism survive in a certain environment is a(n) _____.

4. An adaptation in which an animal is protected against predators by its resemblance to an unpleasant or dangerous animal is _____.

5. A type of camouflage in which the color of an animal blends in with the animal's background is protective _____.

68 Chapter 3 • Interactions in Ecosystems
Reading and Writing

Use with Lesson 3
Adaptation and Survival

Adaptation and Survival

Fill in the blanks.

cactus	poisons	streamlined
camouflage	prey	water
mimicry	seasons	

Both plants and animals have adaptations that help them survive in their environments. For example, plants such as the __cactus__ have thick, waxy stems that conserve __water__ in environments that are hot and dry. Plants in cold climates have shortened growing __seasons__. Ocean animals are more __streamlined__ than land animals so that they can swim faster.

Some adaptations developed because of predator-__prey__ relationships. Plants, such as milkweed, contain __poisons__ that make predators avoid them. Prey can use __camouflage__ to blend in with their environments. Some animals also demonstrate __mimicry__, the ability to look like another animal that a predator finds unpleasant. For example, some predators stay away from the viceroy butterfly because it mimics the bad-tasting monarch butterfly.

Reading in Science

Name _____ Date _____

Meet Caroline Chaboo

Read the Reading in Science feature in your textbook.

Complete the statements in the "Clues" and "What I Know" columns. Use this information to infer something that is not directly stated in the text. Write that statement in the "What I Infer" column.

Clues	What I Know	What I Infer
1. The Sabal palm stands up to high winds, drought, and driving rain in the _____ region.	The Sabal palm is well adapted for the Caribbean region.	
2. The _____ beetle harms Sabal palm trees in regions where it lives.	The tortoise beetle lives in _____.	
3. The tortoise beetle weakens the Sabal palm, but _____.	Caroline Chaboo studies plants, such as the Sabal palm, to discover whether they have adapted natural protection against insect pests.	

Chapter 3 • Interactions in Ecosystems
Reading and Writing

Use with **Lesson 3**
Adaptation and Survival

Name _____ Date _____

Reading in Science

Write About It

Infer Read the "Write About It" questions carefully. Use the text within "Meet Caroline Chaboo" to write your answers.

Using Ideas to Infer

To answer Question #1, first determine how a natural pesticide inside the Sabal palm would help the tree.

_____

_____

Then, write your answer to the question:

How might a natural pesticide in the Sabal palm help other organisms?

_____

_____

Planning and Organizing

Imagine that you have been told to research tortoise beetles to find out what other plants they eat.

In order to conduct this research, first list the types of sources that would contain this information.

a. _____

b. _____

c. _____

Then, list key words you could use to look up the information in these sources.

a. _____

b. _____

Chapter 3 • Interactions in Ecosystems
Reading and Writing

Use with **Lesson 3**
Adaptation and Survival

Interactions in Ecosystems

Choose the letter of the best answer.

1. All the living and nonliving things in an environment make up a(n)
   a. community.
   b. ecosystem.
   c. population.
   d. species.

2. The path that energy takes in an ecosystem as it moves from one organism to another is a(n)
   a. producer chain.
   b. energy pyramid.
   c. food chain.
   d. energy web.

3. Which animal hunts other animals for food?
   a. predator
   b. producer
   c. prey
   d. herbivore

4. A diagram that shows the amount of energy available at each level of an ecosystem is a(n)
   a. energy pyramid.
   b. food web.
   c. food chain.
   d. ecosystem diagram.

5. All of the members of one species in an ecosystem are a(n)
   a. community.
   b. food chain.
   c. environment.
   d. population.

6. Which type of resource restricts population growth within an ecosystem?
   a. biotic factor
   b. limiting factor
   c. capacity factor
   d. energy factor

7. The measure of the size of a population and the area that can support it is
   a. limiting factor.
   b. ecosystem limit.
   c. carrying capacity.
   d. community.

Name _____ Date _____

CHAPTER Vocabulary

Choose the letter of the best answer.

8. A type of symbiosis in which both organisms benefit is
   a. commensalism.
   b. parasitism.
   c. predatorism.
   d. mutualism.

9. A type of symbiosis in which one organism benefits and the other is not harmed is
   a. commensalism.
   b. parasitism.
   c. predatorism.
   d. mutualism.

10. What is the name of the physical place in which an organism lives and hunts for food?
    a. habitat
    b. niche
    c. host
    d. community

11. Any characteristic that helps an organism survive in its environment is a(n)
    a. niche.
    b. mimicry.
    c. adaptation.
    d. abiotic factor.

12. A color, shape, or pattern that allows an organism to blend in with its environment is called
    a. mimicry.
    b. symbiosis.
    c. mutualism.
    d. camouflage.

13. A type of camouflage in which the color of an animal blends with its background is
    a. protective resemblance.
    b. commensalism.
    c. protective coloration.
    d. adaptive coloring.

14. An adaptation in which an animal is protected by it's resemblance to an unpleasant animal that predators avoid is called
    a. camouflage.
    b. symbiosis.
    c. mimicry.
    d. parasitism.

15. What word refers to the special role that an organism plays in a community?
    a. symbiosis
    b. niche
    c. habitat
    d. mutualism

Chapter 3 • Interactions in Ecosystems
Reading and Writing

CHAPTER Concept Map

Name _____ Date _____

Ecosystems and Biomes

Complete the concept map with information you learned about ecosystems and biomes.

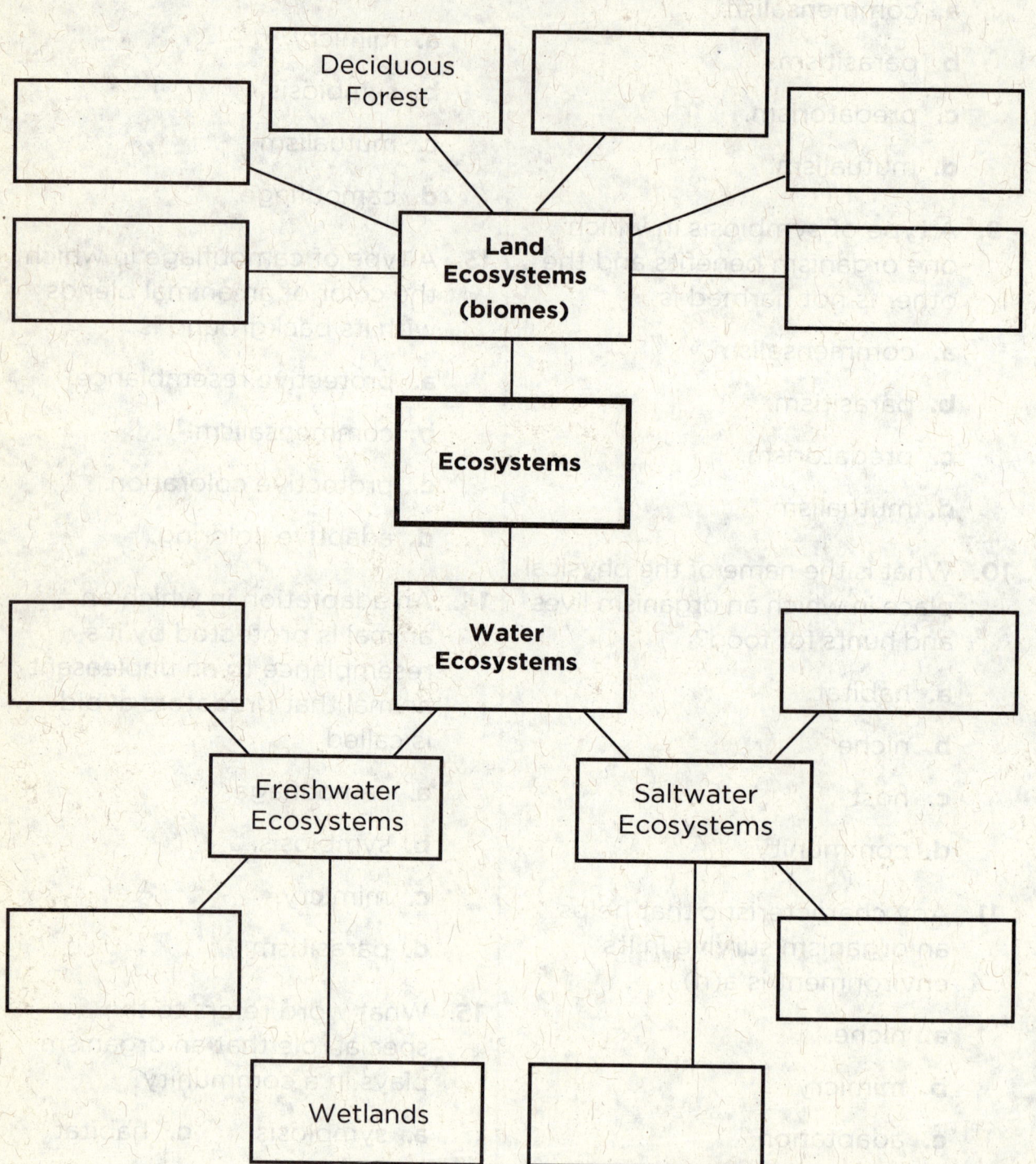

74 Chapter 4 • Ecosystems and Biomes
Reading and Writing

Name _____ Date _____

LESSON Outline

Cycles in Ecosystems

Use your textbook to help you fill in the blanks.

What is the water cycle?

1. The constant movement of water as a liquid and a gas between Earth's air and its surface is the _____. The energy for the water cycle comes from the _____.

2. Solar energy causes water on Earth's surface to change to gas and rise into the atmosphere during the process of _____.

3. As gaseous water rises and cools, it turns into droplets of liquid water during the process of _____.

4. Water droplets may fall from the atmosphere back to Earth's surface as _____, such as rain and snow.

5. Some water soaks into the ground and collects in pores in soil and rock to become _____.

6. Plants take water from the soil and return it to the air through their leaves during _____.

What is the carbon cycle?

7. The constant exchange of carbon among Earth's living organisms is the _____.

8. During photosynthesis, plants take _____ from the air and use it to make food that other living things use.

Chapter 4 • Ecosystems and Biomes
Reading and Writing

Use with **Lesson 1**
Cycles in Ecosystems

Lesson Outline

Name _____ Date _____

9. Animals and plants burn carbon-rich foods during cellular _____, and then release carbon dioxide to the atmosphere.

10. Fossils fuels, which form from the decayed remains of living things, release the carbon in them when they are _____.

What is the nitrogen cycle?

11. Although most of Earth's atmosphere is nitrogen, the gas must first be _____ so that it is in a form that most living things can use.

12. Nitrogen can be fixed by volcanic action, by _____, and by nitrogen-fixing _____.

13. Plants use nitrogen to make _____, which get into the bodies of animals when they eat plants or plant-eating animals.

14. Nitrogen returns to the soil in the _____ of animals, and when plants and animals decay.

How is matter recycled?

15. People can reduce their use of natural resources through _____.

Critical Thinking

16. Why is it necessary for water, carbon, nitrogen, and oxygen to be recycled through Earth's ecosystems?

_____

_____

Name _____ Date _____

LESSON Vocabulary

Cycles in Ecosystems

What am I?

Choose a word from the word box below that answers each question.

a. carbon cycle	**d.** evaporation	**g.** precipitation
b. compost	**e.** groundwater	**h.** runoff
c. condensation	**f.** nitrogen cycle	

1. _____ I am the changing of a gas into a liquid as the gas cools. What am I?

2. _____ I am the continuous changing of nitrogen gas into compounds in the soil and its later release back to the air. What am I?

3. _____ I am precipitation that flows over the land's surface into rivers and lakes and is not absorbed into the ground. What am I?

4. _____ I am a mixture of decayed plant and animal material that can be used as fertilizer. What am I?

5. _____ I am the continuous exchange of carbon among living things. What am I?

6. _____ I am the changing of a liquid into a gas. What am I?

7. _____ I am water that sinks beneath the ground and collects in tiny holes in soil and rock. What am I?

8. _____ I fall to Earth's surface as rain, snow, sleet, or hail. What am I?

Chapter 4 • Ecosystems and Biomes
Reading and Writing

Use with **Lesson 1
Cycles in Ecosystems**

LESSON Cloze Activity

Name _____ Date _____

Cycles in Ecosystems

Fill in the blanks.

atmosphere	fixation	runoff
decomposers	nitrates	
evaporation	precipitation	

The Sun provides energy for the water cycle. Heat causes water to rise from Earth's surface as a gas through a process called _____ . Water then condenses in the atmosphere and falls, as _____ , into oceans and the ground, or over land as _____ .

In the carbon cycle, plants use carbon dioxide to make food. Carbon dioxide is released back into the _____ when living things use this food. Organisms called _____ release carbon dioxide when they break down dead organisms.

Nitrogen _____ places nitrogen in a form in which it can be used by most living organisms. Plants use nitrogen in the form of _____ . Other organisms get nitrogen from plants and return it to the soil through their wastes.

Name _____ Date _____

Changes in Ecosystems

Use your textbook to help you fill in the blanks.

How can ecosystems change?

1. Ecosystems are changed by living _____ that change the environment around them, and by _____ events such as floods.

2. Humans can change or destroy the _____ of organisms when they cut _____ to build homes.

What happens when ecosystems change?

3. Some organisms respond to changes in ecosystems by adapting or _____ to another place.

4. When a type of organism cannot respond to changes in an ecosystem, it may become _____ .

5. When a species is in danger of extinction, it is called an _____ species.

6. Species that could become endangered because of their low populations are known as _____ species. The biggest threat to a species is the loss of _____ .

How do ecosystems come back?

7. Over time, one group of species in an ecosystem is replaced by a different group of species through a process called _____ .

Chapter 4 • Ecosystems and Biomes
Reading and Writing

Use with **Lesson 2**
Changes in Ecosystems

Lesson Outline

Name _____ Date _____

8. In a region where few if any species existed before or where previous species were wiped out, _____ occurs.

9. The first species to take hold in barren areas are _____ species, such as mosses and lichens.

10. As larger plants and predators begin to live in an area, the community may become a(n) _____, such as a prairie.

11. With enough moisture, _____ may start to grow in a grassland.

12. In time, a fully developed ecosystem will support a(n) _____ community, which is the final stage of succession.

What is secondary succession?

13. When a new community develops where a community had once existed, it is called _____ succession.

14. Secondary succession might occur in a forest that has been burned by a(n) _____ or at an abandoned farm.

Critical Thinking

15. A volcano erupts and lava flows over what had once been a fertile farm field. Describe the type of succession that will occur, and explain why.

_____

_____

Name _____ Date _____

LESSON Vocabulary

Changes in Ecosystems

Match the correct letter with the description.

> **a.** climax community
> **b.** endangered species
> **c.** extinct
> **d.** pioneer community
> **e.** pioneer species
> **f.** primary succession
> **g.** secondary succession
> **h.** succession

1. Establishment of the first living community to develop in an area that used to be lifeless is called _____ .

2. When a species dies out completely, the species is _____ .

3. The establishment of a new community where a community had already existed is called _____ .

4. The process of one ecosystem gradually changing into a different type of ecosystem is called _____ .

5. A species that is in danger of becoming extinct is a(n) _____ .

6. In the final stages of succession, a(n) _____ develops.

7. One of the first species to live in an area that used to be lifeless is a(n) _____ .

8. Succession that occurs where there is no soil and where few, if any, living things exist is _____ .

Chapter 4 • Ecosystems and Biomes
Reading and Writing

Use with Lesson 2
Changes in Ecosystems

LESSON Cloze Activity

Name _____ Date _____

Changes in Ecosystems

Fill in the blanks.

animal	plants	species
habitat	primary succession	trees
pioneer	secondary succession	

Ecosystems change over time. People cause some of the changes, through pollution, _____ destruction, or hunting, or by introducing or removing _____.

However, many ecosystem changes are natural. When land is burned by a fire or a farm field is abandoned, _____ occurs. New _____ begin to grow in the soil. Weeds, then shrubs, and finally _____ grow. When few, if any, living things exist in an area, _____ will establish a first community. The first organisms to live in the area are called _____ species. After soil is established, larger plants can grow and larger _____ species can arrive. Eventually, forests develop. Finally, in the last stage of succession, a climax community is established.

82 Chapter 4 • Ecosystems and Biomes
Reading and Writing

Use with **Lesson 2**
Changes in Ecosystems

Name _____ Date _____

LESSON Outline

Biomes

Use your textbook to help you fill in the blanks.

What are biomes?

1. Each of Earth's major land ecosystems is a(n) _____ . Each biome has its own specific animals, plants, soil, and _____ .

2. A sandy or rocky biome with a dry climate is a(n) _____ . Some organisms have _____ that allow them to survive in dry regions.

What are some harsh biomes?

3. The ground in the _____ stays frozen all year. Trees cannot grow where this layer of constantly frozen ground called _____ exists.

4. Some grasses, _____ , and _____ grow in the tundra.

5. Although few animals live in the tundra, _____ bears, caribou, and Arctic _____ do make their homes there.

6. The _____ is a cool, forest biome just south of the tundra.

7. The dominant type of vegetation in the taiga biome is _____ .

8. Many of the animals in the taiga have thick _____ and layers of fat to protect them from cold weather.

Chapter 4 • Ecosystems and Biomes
Reading and Writing

Use with **Lesson 3** Biomes 83

LESSON Outline Name _____ Date _____

What are some forest biomes?

9. A hot biome near the equator that has lots of rain and more plants and animals than any other biome is the _____.

10. This biome has four _____, with different plants and animals in each one.

11. The _____ rain forest biome has lots of rain and a cooler climate than tropical forests.

12. The _____ is a forest biome with four seasons and trees that lose their leaves in autumn.

13. Winter in the deciduous forest can be cold, and many animals hibernate, _____ to warmer climates.

What are grasslands?

14. The _____ is a biome where grasses, not trees, are the main type of plant life. In North America, the _____ is a large area of grassland.

15. The grassland biome is wetter than that of a desert but does not have enough precipitation to support many _____.

Critical Thinking

16. Why is climate important in determining biomes?

_____
_____
_____
_____

Name _____ Date _____

LESSON Vocabulary

Biomes

Use the clues below to help you fill in the blanks.

| biome | desert | taiga | tropical |
| deciduous | grassland | temperate | tundra |

1. The _____ is a large, treeless biome where the ground is frozen all year.

2. A very rainy biome called the _____ rain forest is dominated by evergreen trees and has mild winters and cool summers.

3. Any of Earth's major land ecosystems with its own typical plants, soil, and climate is a(n) _____ .

4. The _____ is a cool, northern forest biome dominated by conifers.

5. The _____ forest, a forest biome with four distinct seasons, has trees that lose their leaves each year in autumn.

6. The _____ is a sandy or rocky biome that has little precipitation and limited plant life.

7. With few trees, the _____ is a biome in which the main form of vegetation is grass.

8. The _____ rain forest is a hot, humid biome near the equator, that has abundant rainfall and a wide variety of life.

Chapter 4 • Ecosystems and Biomes
Reading and Writing

Use with Lesson 3
Biomes

LESSON Cloze Activity

Name _____ Date _____

Biomes

Fill in the blanks.

climate	hardwood	taiga
deciduous forest	permafrost	temperate
grassland	rainy	tropical rain forest

Earth has several major land ecosystems called biomes. Each of these has its own typical animals, plants, soil, and _____. The _____ biome is hot and _____ all the time and has more types of plants and animals than any other biome. There are also _____ rain forests, which are rainy, but have a cooler climate than tropical forests. In the _____ biome, deciduous trees dominate. These are _____ trees that lose their leaves each autumn. North of this biome is the _____, with its cold, snowy climate and forests of conifers.

The coldest, harshest biome is the tundra, which is a treeless area with a layer of _____ under the surface. Another largely treeless biome is the _____, where grasses are the main type of plant life. The desert biome is sandy or rocky, with little precipitation or plant life.

Name _____ Date _____

Reading in Science

Did You Know That Forests Breathe?

Read the passage titled "A Year in the Life of a Forest" in your textbook. The passage about the Howland Forest of Maine contains five paragraphs. In the blanks provided in the graphic organizer, write a sentence that summarizes the main idea of the first three paragraphs, followed by two sentences that contain supporting details. Use your own words. The first item has been done for you.

Main Idea	Details
Paragraph 1 Main Idea: Scientists measure gas levels in forests throughout the year.	Howland Forest is a deciduous forest in Maine.
	The change in seasons affects the levels or carbon dioxide there.
Paragraph 2 Main Idea:	
Paragraph 3 Main Idea:	

Chapter 4 • Ecosystems and Biomes
Reading and Writing

Use with Lesson 3
Biomes

87

Reading in Science Name _____ Date _____

Write About It

Main Idea and Details 1. Tell how the levels of carbon dioxide change in the Howland Forest throughout the year. 2. Research other biomes, and explain how they change during the year.

Now, use the information in your graphic organizer to write a paragraph telling how the levels of carbon dioxide change in the Howland Forest throughout the year.

_____
_____
_____
_____
_____
_____
_____
_____
_____

Next, you will be conducting research about the yearly changes to another biome.

1. What biome do you choose to research? _____

2. What types of organisms live in this biome? _____

3. What changes can be observed in this biome as the seasons change? _____

4. Compare your biome research with that of the students seated closest to you. Why do seasonal changes in the different biomes vary? Give your opinion.

Name _____ Date _____

LESSON Outline

Water Ecosystems

Use your textbook to help you fill in the blanks.

What are water ecosystems?

1. There are freshwater ecosystems and _____ ecosystems.

2. Organisms that drift in the water are called _____ . Active swimmers such as fish are called _____ .

3. The creatures that live in the deepest part of a body of water are the _____ . Many bottom-living creatures are scavengers or _____ .

4. Producers, which live at or near the surface, release the _____ that allows most other water organisms to live in surface waters.

What are freshwater ecosystems?

5. Organisms in running-water ecosystems are adapted to how _____ the water flows.

6. In standing-water ecosystems, such as lakes, most organisms live in the shallow water of the _____ zone.

7. Many nekton live in the _____ zone, which is away from the shore.

8. Benthos, including worms and mollusks, live in the _____ zone beneath the open-water zone.

Chapter 4 • Ecosystems and Biomes
Reading and Writing

Use with **Lesson 4**
Water Ecosystems

LESSON Outline

Name _____ Date _____

What are ocean ecosystems?

9. Organisms of the shallow _____ zone are covered and uncovered each day by the rise and fall of tides.

10. Sunlight allows producers and the animals that depend on them to live in the _____ zone.

11. Large organisms live near the surface in the top part of the _____ zone, which is called the bathyal zone.

12. Few creatures can live in the cold, dark waters at the bottom of the oceanic zone, which is called the _____ zone.

Where do salt and fresh water meet?

13. The place where a river empties into the ocean is called a(n) _____ . Estuaries usually contain _____ marshes, boggy areas covered with grasses.

14. When the tide comes in, an estuary's waters are mostly _____ , but the waters are mostly _____ when the tide goes out.

15. Wetlands protect coastal regions during _____ by soaking up excess water.

Critical Thinking

16. How is sunlight a limiting factor in water ecosystems?

_____
_____
_____

Name _____ Date _____

Lesson Vocabulary

Water Ecosystems

Match the correct letter with the description and fill in the crossword puzzle.

| benthos | nekton | shore zone |
| intertidal zone | plankton | |

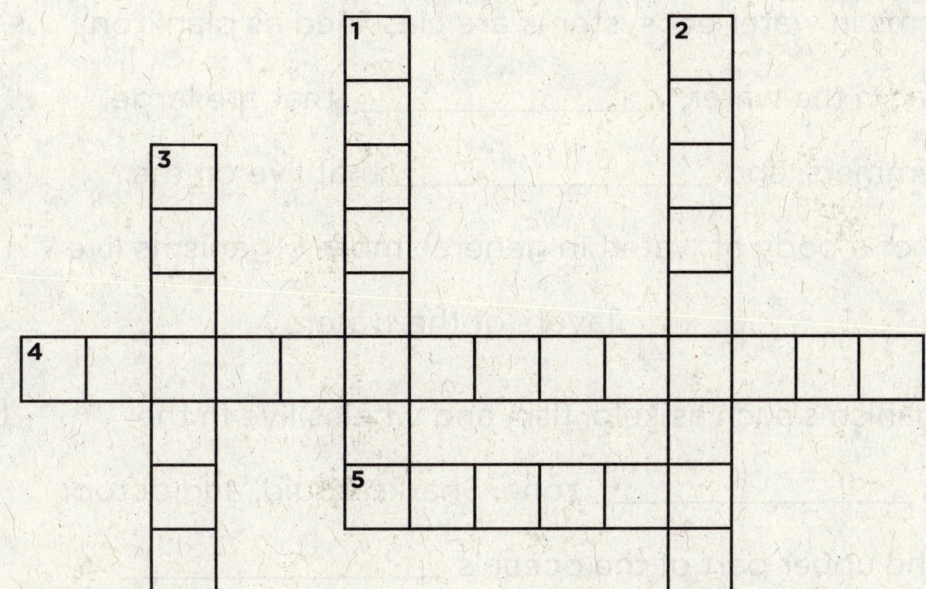

Across

4. Place where organisms are covered and uncovered daily by the waters of changing tides

5. The larger, active swimmers in a body of water

Down

1. Creatures that drift freely in the water

2. The shallow water in standing-water ecosystems

3. Organisms that live on the bottom of a body of water

Chapter 4 • Ecosystems and Biomes
Reading and Writing

Use with Lesson 4
Water Ecosystems

Water Ecosystems

Fill in the blanks.

| benthos | nekton | oceanic | tides |
| intertidal | neritic | running-water | upper |

Water ecosystems have many forms of life. Organisms in water ecosystems are classified as plankton that float in the water; _____ that are large, free swimmers; and _____ that live on the bottom of a body of water. In general, more organisms live in the _____ layers of the water.

Organisms such as kelp, fish, and whales live in the ocean's _____ zone. Sharks, squid, and octopi live in the upper part of the ocean's _____ zone (few animals live in the lower part of this zone).

Freshwater ecosystems are divided into _____ bodies, standing-water bodies, and wetlands ecosystems.

Organisms of the ocean's (saltwater) _____ zone must be adapted to rise and fall of _____. Organisms that live in estuaries are adapted to survive in both fresh and salty waters.

Name _____ Date _____

Writing in Science

Keep Our Water Clean

Write About It
Write a letter to the mayor of your town. Explain a need that the students in your community have and why people should help. State your opinion clearly and support it with relevant facts and evidence organized in a logical way.

Getting Ideas

Think of an issue that clearly affects life in your community. Form an opinion about it. Write this opinion in the top box in the chart below. Then jot down reasons that support this opinion in the bottom boxes.

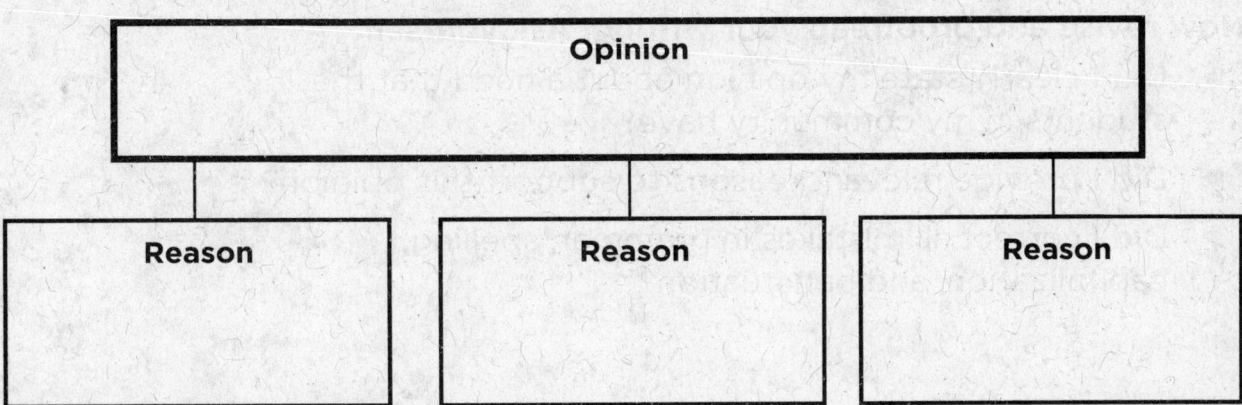

Planning and Organizing

Kristin's opinion is that the town should ban power boats from the lake. Here are three statements she wrote. Write Yes if the statement supports her opinion. Write No if it does not.

1. _____ Power boats can pollute the water.

2. _____ Power boats are a fun way to spend time on the lake.

3. _____ Power boats kill fish and other water life.

Chapter 4 • Ecosystems and Biomes
Reading and Writing

Use with **Lesson 4**
Water Ecosystems

Writing in Science

Drafting

Write a statement to begin your letter. Make sure it clearly tells the issue you are concerned about and states your opinion.

_____

_____

_____

Now write your letter to the mayor on a separate piece of paper. Use the form of a business letter. Start the body of your letter with the sentence you wrote above. Include reasons that will persuade the mayor to support your opinion. End by stating what you think should be done. Remember to be polite and respectful.

Revising and Proofreading

Now revise and proofread your writing. Ask yourself:

▶ Did I clearly state my opinion about a need that the students in my community have?

▶ Did I provide relevant reasons to support this opinion?

▶ Did I correct all mistakes in grammar, spelling, capitalization, and punctuation?

Name _____ Date _____ **CHAPTER Vocabulary**

Ecosystems and Biomes

Choose the letter of the best answer.

1. Which of these is a type of precipitation?
   a. frost
   b. clouds
   c. dew
   d. hail

2. In which natural cycle must an important gas in Earth's atmosphere be fixed before plants can use it?
   a. sulfur cycle
   b. carbon cycle
   c. nitrogen cycle
   d. oxygen cycle

3. Which of these processes is the changing of water vapor into liquid water?
   a. condensation
   b. infiltration
   c. evaporation
   d. transpiration

4. Which kind of species is in danger of totally disappearing from Earth?
   a. threatened
   b. endangered
   c. extinct
   d. pioneer

5. In the last stage of succession, the plants and animals in an ecosystem form a(n)
   a. pioneer community.
   b. endangered community.
   c. climax community.
   d. primary community.

6. Which type of succession would occur after a fire has burned a forest?
   a. primary succession
   b. tertiary succession
   c. secondary succession
   d. climax succession

Chapter 4 • Ecosystems and Biomes
Reading and Writing

CHAPTER Vocabulary

Choose the letter of the best answer.

7. Which type of biome has the greatest diversity of plants and animals?
   a. tundra
   b. deciduous forest
   c. desert
   d. tropical rain forest

8. In which cold, northern biome are conifers the main type of plant life?
   a. deciduous forest
   b. tropical rain forest
   c. taiga
   d. tundra

9. In which biome do hardwood trees lose their leaves before the cold winter sets in?
   a. tundra
   b. taiga
   c. deciduous forest
   d. tropical rain forest

10. The cold temperatures and frozen ground prevent the growth of trees in the
    a. taiga.
    b. desert.
    c. tundra.
    d. grasslands.

11. Creatures that drift freely in water ecosystems are called
    a. plankton.
    b. benthos.
    c. nekton.
    d. crustaceans.

12. What are the large, active swimmers, such as fish and whales, in water ecosystems?
    a. mollusks   c. plankton
    b. benthos   d. nekton

13. The organisms that live along the bottom of water ecosystems are the
    a. benthos.   c. plankton.
    b. nekton.   d. shellfish.

The Many Sides of Diamonds

Read the Literature feature in your textbook.

Write About It

Response to Literature This article describes the formation and use of diamonds. Research additional information about the history of industrial diamonds, how they are formed, and how they are used. Write a report about industrial diamonds. Include facts and details from this article and from your research.

CHAPTER Concept Map

Our Dynamic Earth

Complete the concept map by filling in answers where blanks appear.

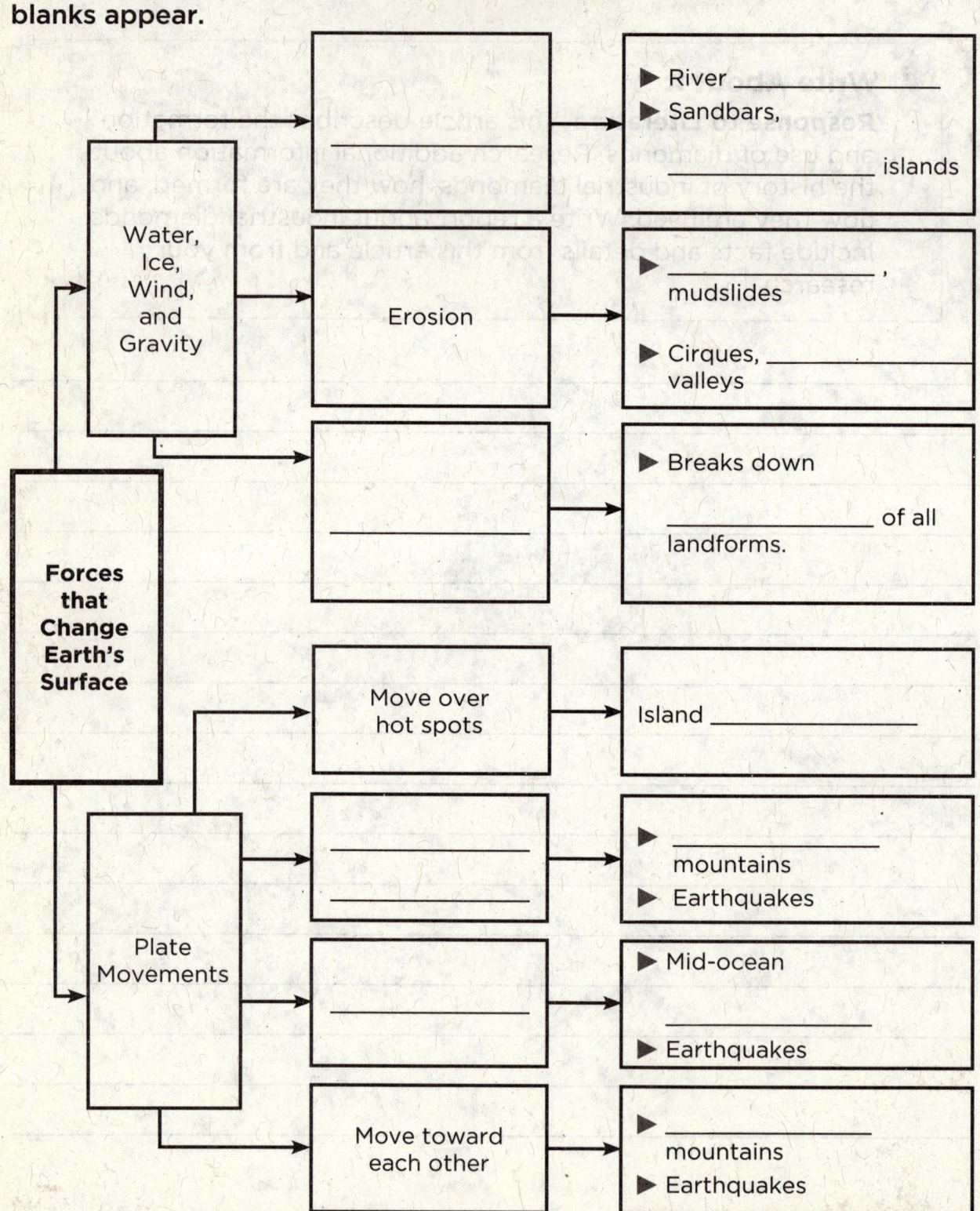

98 Chapter 5 • Our Dynamic Earth
Reading and Writing

Name _____ Date _____

LESSON Outline

Earth's Landforms

Use your textbook to help you fill in the blanks.

What are landforms?

1. A physical feature on Earth's surface is a(n) _____.

2. The highest of Earth's physical features are _____.

3. A low area between mountains or hills is a(n) _____.

4. Wide, flat areas of land are called _____.

5. A large, flat area higher than the land around it is a(n) _____.

6. Earth's largest bodies of water are its saltwater _____.

7. Natural streams of flowing water that empty into lakes, oceans, or other bodies of water are _____.

8. A body of water with land all around it is a(n) _____.

What are the features of the ocean floor?

9. A(n) _____ is a large underwater area between continents.

10. Shallow waters cover the _____, the gently sloping part of the ocean floor along the coast.

11. The sharp drop from the continental shelf to the continental rise is the _____.

Chapter 5 • Our Dynamic Earth
Reading and Writing

Use with Lesson 1
Earth's Landforms

99

LESSON Outline   Name _____ Date _____

12. A wide, flat area covering about 40 percent of the ocean floor is the _____ .

13. The deepest areas of the ocean floor are _____ .

How are Earth's features mapped?

14. Measurements taken by a(n) _____ are used to make maps.

15. Elevations are shown with shading on a(n) _____ map.

16. Lines are used to show elevation and steepness of slopes on a(n) _____ map.

What are Earth's layers?

17. The layer of air around Earth is the _____ .

18. Earth's waters make up Earth's _____ .

19. Earth is made of three main layers: the crust, the _____ , and the core.

20. The part of Earth that is home for living things is the _____ .

Critical Thinking

21. Compare the mantle and core of the Earth.

_____
_____
_____
_____

100 Chapter 5 • Our Dynamic Earth
Reading and Writing

Use with **Lesson 1**
Earth's Landforms

Name _____ Date _____

LESSON Vocabulary

Earth's Landforms

Match the correct word with its description, and fill in the crossword puzzle.

> atmosphere   crust   landform   mantle
> core   hydrosphere   lithosphere

Across

3. _____ a physical feature on Earth's surface

5. _____ the layer of air that surrounds Earth

6. _____ the central part of Earth

Down

1. _____ formed by Earth's liquid and solid water

2. _____ the rocky upper layer of Earth that contains continents and ocean basins

3. _____ the crust and the top of the upper mantle form it

4. _____ the layer of Earth's interior below the crust

Earth's Landforms

Fill in the blanks.

crust	inner core	oceans
elevation	landforms	outer core
hydrosphere	mantle	surveyor

The physical features of Earth are part of Earth's surface. Earth's surface has many types of _____, from high mountains to deep valleys. There are also physical features under Earth's largest bodies of water, the _____. These undersea features look like the mountains, valleys, and cliffs on land.

Landforms are measured by a(n) _____. One important measurement is _____, or the height of land above sea level.

Earth has several layers. Earth's waters are its _____. The planet itself is divided into the surface _____, the _____ beneath it, and the core at the center. The crust and mantle are rock. Earth's core is made of metal. The _____ is liquid, and the _____ is solid. The core makes up the central part of Earth.

Name _____ Date _____

LESSON Outline

Plate Tectonics

Use your textbook to help you fill in the blanks.

Is Earth's crust moving?

1. Geologist Alfred Wegener formulated the theory of _____.

2. Wegener's theory stated that Earth's _____ were once joined in one landmass, but gradually pulled apart and drifted.

3. Wegener's showed that the age and composition rocks in the _____ on South America's east coast matched of those on Africa's west coast.

4. Scientists also discovered evidence in _____ that Africa and South America were once joined.

How does the movement of Earth's crust affect the ocean?

5. Scientists developed the _____ model to explain how the continents have moved over millions of years.

6. Earth's lithosphere is made of huge pieces of solid rock called _____.

7. These solid pieces of rock rest on the hot, soft, slippery rock of Earth's _____.

8. Melted rock called _____ rises up through the crack where plates move apart under the ocean.

9. As the ocean floor spreads at the plate boundary, the _____ resting on the plates also move apart.

Chapter 5 • Our Dynamic Earth
Reading and Writing

Use with Lesson 2
Plate Tectonics

LESSON Outline Name _____ Date _____

10. The hot rock cools at the surface forming the mid-ocean ridge and the _____ along its top.

How does the movement of Earth's plates affect the land?

11. When plates push toward each other, a force called _____ results.

12. Because of this force, the ground at the edges of plates is pushed upward to form _____ mountains.

13. A mountain range in Asia, the _____, began to form in this way millions of years ago.

14. In places where one plate rubs past another, a twisting or tearing force called _____ results.

15. This force can cause blocks of crust to break apart along deep cracks in Earth's crust called _____.

16. When rock on one side of a fault moves down and rock on the other side moves up, a _____ mountain is formed.

17. A California mountain range, the _____, is this type of landform.

Critical Thinking

18. Compare how two types of mountains are formed.

_____
_____
_____
_____

Name _____ Date _____

LESSON Vocabulary

Plate Tectonics

Use the terms in the box below to fill in the blanks.

continental drift	mid-ocean ridge
fault-block mountains	plate tectonics
folded mountains	rift valley
geologist	

1. As hardened magma builds up on both sides of a plate boundary, a(n) _____ forms at the center of a mid-ocean ridge.

2. The force of compression can form _____ at the point where two plates push together.

3. The theory that states that the continents were once one landmass and that they drifted to their present positions over many years is called _____ .

4. A highland in the middle of the oceans that runs parallel to the continents is called a(n) _____ .

5. A scientist who studies Earth's structure and history is called a(n) _____ .

6. Shear forces at a fault can form _____ .

7. The scientific theory that states Earth's crust is made of moving plates is called _____ .

LESSON Cloze Activity

Name _____ Date _____

Plate Tectonics

Fill in the blanks.

| compression | continents | fossils | shear |
| continental drift | folded | plate tectonics | |

The continents were not always where they are today. About 100 years ago, Alfred Wegener developed the theory of _____. The theory states that Earth's _____ were once one landmass. The landmass broke up millions of years ago, and the continents drifted to the positions we know today. Wegener supported his theory with evidence from rocks and _____. Later, scientists developed the theory of _____.

When plates push together, they produce the force of _____. This force can push the ground at the boundary upward, forming _____ mountains. When plates slide past each other, they create _____. This force can make huge blocks of crust break apart along faults. Over millions of years, the blocks can shift upward to form fault-block mountains.

Name _____ Date _____

Reading in Science

Pangea and Other Superconstituents

Write About It
What evidence do scientists have that Rodinia and Pannotia existed? Research this evidence and select a main idea. Write an expository essay with details to support your main idea.

Getting Ideas
Do some research to find out whether Rodinia and Pannotia actually existed. Use the chart below. In the boxes on the top, write details that you find. In the box on the bottom, summarize this information.

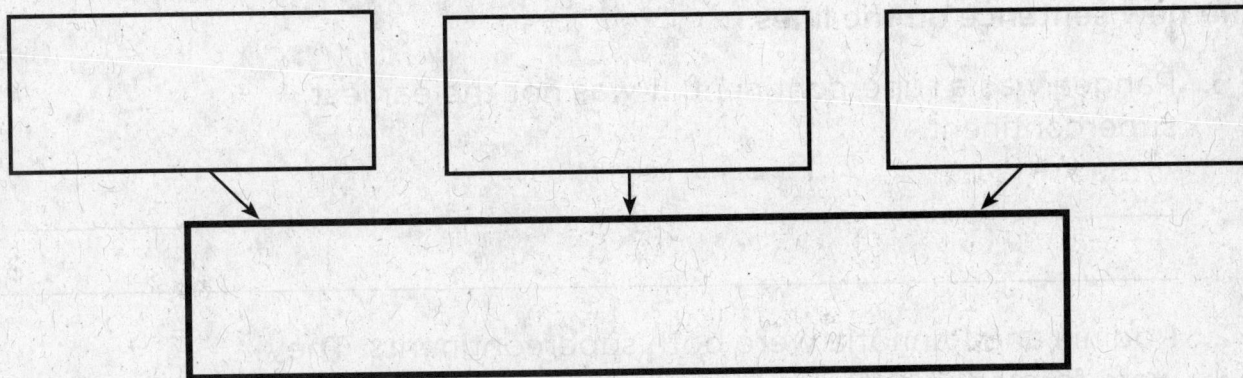

Planning and Organizing
Here are two sentences that Mai wrote. Write Yes if the sentence supports the idea that Rodinia and Pannotia actually existed. Write No if it does not.

1. There are common rock types and structural features along the coastlines of continents today. _____

2. Figuring out how supercontinents formed and broke apart is a lot like detective work. _____

Chapter 5 • Our Dynamic Earth
Reading and Writing

Use with Lesson 2
Plate Tectonics   107

Reading in Science

Name _____ Date _____

Drafting

Write a sentence to begin your essay. This sentence should tell your main idea about Rodinia and Pannotia.

_____
_____
_____

Review the evidence you found and your summary. Now write the first draft of your essay. Use a separate piece of paper. Include facts and details that back up your main idea. Draw a conclusion at the end.

Revising and Proofreading

Help Mai revise her writing. Use the word *but* to combine each pair of sentences. Put a comma before this word. Write the new sentence on the lines.

1. Pangea was a supercontinent. It was not the earliest supercontinent.

    _____
    _____

2. Rodinia and Pannotia were both supercontinents. They were formed at different times.

    _____
    _____

Now revise and proofread your writing. Ask yourself:

▶ Did I clearly state my main idea?
▶ Did I include facts and details to back up my idea?
▶ Did I reach a sound conclusion at the end?
▶ Did I correct all mistakes?

Name _____ Date _____

LESSON Outline

Volcanoes

Use your textbook to help you fill in the blanks.

Where are volcanoes found?

1. Most of Earth's volcanoes are located at edges of _____ .

2. A string of volcanoes at plate boundaries around the Pacific Ocean is known as the _____ .

3. Volcanoes often erupt at places where one plate _____ the other.

4. The bottom edge of the diving plate melts in the heat of the _____ .

5. The melted rock rises within the crust, forming a hot pool of _____ .

6. The hot rock sometimes erupts through openings in Earth's surface as a(n) _____ .

7. Magma that reaches Earth's surface is _____ .

How do volcanoes build land?

8. When magma hardens inside Earth's crust, it can form vertical _____ and horizontal sills.

9. Magma pushed into a thick sill can form a(n) _____ .

10. The largest underground magma formations are _____ , which can form large hills.

Chapter 5 • Our Dynamic Earth
Reading and Writing

Use with **Lesson 3** Volcanoes

LESSON Outline Name _____ Date _____

11. A volcano that is _____ can erupt with lava, ash, gas, or rock.

12. When a volcano stays quiet for a time, it is _____.

13. A volcano that no longer erupts is _____, or dead.

How do volcanoes build islands?

14. The Hawaiian Islands formed over a stationary pool of magma below Earth's crust called a(n) _____.

15. When the mountains grew high enough to break the ocean's surface, they became volcanic _____.

16. As the plate moved away from the hot spot a new _____ formed.

17. Where two ocean plates meet and one is pushed under the other, an island _____ may form.

18. Magma from edge of the lower plate rises and builds volcanic islands along the plate _____.

19. An example of an island arc is the _____ in Alaska.

Critical Thinking

20. Why do volcanoes form when one plate dives under another?

_____

_____

_____

Chapter 5 • Our Dynamic Earth
Reading and Writing

Use with **Lesson 3**
Volcanoes

Name _____ Date _____

LESSON Vocabulary

Volcanoes

Match the correct letter with the description.

> **a.** cinder-cone volcano **f.** island chain
> **b.** composite volcano **g.** lava
> **c.** crater **h.** shield volcano
> **d.** hot spot **i.** volcano
> **e.** island arc

1. _____ magma that reaches Earth's surface

2. _____ a series of volcanic islands that form along a plate boundary

3. _____ a broad volcano with gently sloping sides formed from thin, fluid lava

4. _____ an opening in Earth's crust through which magma flows

5. _____ a stationary pool of magma below Earth's crust

6. _____ a large, cone-shaped volcano built from alternating layers of cinders and hardened lava

7. _____ a line of islands

8. _____ a cup-shaped depression that forms around a volcano's vent

9. _____ a cone-shaped volcano of cinders, with a narrow base and steep sides

Chapter 5 • Our Dynamic Earth
Reading and Writing

Use with **Lesson 3**
Volcanoes

LESSON Cloze Activity

Volcanoes

Fill in the blanks.

| cinder-cone | lava | plates | volcano |
| composite | mantle | shield | |

Openings on Earth's surface appear on the edges of the crust's plates. An opening in Earth's crust from which magma flows is a(n) _____. Most volcanoes form in places where _____ push toward each other, and one dives under the other. The lower edge of the diving plate melts in the _____, producing hot magma that rises in the crust. Magma that breaks through to Earth's surface is _____.

There are three types of volcanic mountains. A large, broad mountain composed of hardened lava is a(n) _____ volcano. A narrow, steep mountain formed from cinders is a(n) _____ volcano. A large, cone-shaped mountain formed by layers of ash and cinders sandwiched between layers of hardened lava is a(n) _____ volcano. Volcanoes are built up over time as more material is deposited.

Name _____ Date _____

LESSON Outline

Earthquakes

Use your textbook to help you fill in the blanks.

What is an earthquake?

1. Earthquakes happen when the layers of rock on both sides of a(n) _____ suddenly slip.

2. Waves of energy spread out from the _____, the place where the slipping began.

3. When they reach the surface, waves spread out from the _____ of the earthquake (the point directly above the focus).

4. Most earthquakes happen at faults that are near the boundaries of _____.

What waves do earthquakes make?

5. Scientists use a(n) _____ to detect and measure earthquake waves.

6. The fastest earthquake waves, _____ waves, pass through solids and liquids and move back and forth.

7. An earthquake's _____ waves travel slower than primary waves and move only through Earth's solid layers.

8. The slowest-moving waves, _____ waves, move across Earth's surface causing the most damage.

How is an earthquake's energy measured?

9. A measure of the amount of _____ that an earthquake releases is magnitude.

10. Scientists use the _____ Scale to measure earthquake magnitude.

Chapter 5 • Our Dynamic Earth
Reading and Writing

Use with Lesson 4
Earthquakes

LESSON Outline Name _____ Date _____

11. Scientists use the _____ Scale to measure an earthquake's effects.

12. An underwater earthquake can produce a large wave called a(n) _____.

13. Underwater earthquakes with a magnitude of _____ or greater on the Richter scale are most likely to cause tsunamis.

How can people prepare?

14. Layers of rubber and steel between a building and its foundation allow the building to _____, reducing the damage caused by up-and-down motions.

15. Before an earthquake, people should _____ objects to prevent them from falling and causing injury.

16. In their attempt to tell when earthquakes might happen, scientists look for possible warning signs such as changes in the angle of the _____.

17. Earthquakes are hard to _____, but the ability to do so would allow early warnings that could save lives.

Critical Thinking

18. Which scale do you think would better explain an earthquake to you, the Richter Scale or the Mercalli Scale? Why?

_____
_____
_____

Name _____ Date _____

LESSON Vocabulary

Earthquakes

Use the clues below to find the words hidden in the puzzle.

```
S M G A R J F D H O
Z M F O I G A D B Y
E A R T H Q U A K E
C G C D E L L X R T
V N E R G K T I W S
R I C H T E R B O P
G T S U N A M I Y U
V U H I P F O C U S
X D G B E C K L T T
Z E P I C E N T E R
```

1. A sudden movement of Earth's crust is a(n) _____.

2. The point on the surface directly above an earthquake's focus is its _____.

3. A crack in Earth's crust is a(n) _____.

4. The place along a fault where the slipping that causes an earthquake begins is the earthquake's _____.

5. A measure of the energy that an earthquake releases is its _____.

6. A large ocean wave caused by an underwater earthquake is a(n) _____.

7. The scale that measures the magnitude of an earthquake is called the _____ Scale.

Chapter 5 • Our Dynamic Earth
Reading and Writing

Use with Lesson 4
Earthquakes

LESSON Cloze Activity

Name _____ Date _____

Earthquakes

Fill in the blanks.

earthquake	primary or P
energy	Richter
fault	secondary or S
Mercalli	

The plates of the Earth are in motion. A sudden movement of Earth's crust is a(n) _____. Most earthquakes occur near plate boundaries, when layers of rock that usually adhere to each other suddenly slip at a(n) _____. The scale that measures the magnitude of an earthquake is called the _____ Scale. The scale that measures how severe an earthquake feels and the amount of damage the quake does to objects is called the _____ Scale.

The movement of plates during an earthquake sends out waves of _____ that shake the ground. When an earthquake occurs, _____ waves move back and forth very rapidly. An earthquake's _____ waves move up and down. The slowest waves are surface or L waves. They cause the most damage.

Name _____ Date _____

Writing in Science

How Earthquakes Help Predict Volcanic Eruptions

Write About It
What are the differences between short-period and long-period earthquakes? Research these earthquakes. Write an explanatory essay with details that support your main idea.

Getting Ideas

Find out more about these types of earthquakes. Use the chart below to record information. Write the cause of each type of earthquake in the left-hand box. Write the type of earthquake in the right-hand box.

Cause	→	Effect
	→	
	→	
	→	

Planning and Organizing

Hector wants to write first about short-term earthquakes, then about long-term earthquakes, and finally about harmonic-tremor earthquakes. Here are three sentences he wrote. Help him organize them. Then write 1 by the sentence that should come first. Write 2 by the sentence that should come second. Write 3 by the sentence that should come last.

Writing in Science   Name _____ Date _____

1. _____ Seismographs can detect the ongoing flow of magma in a harmonic-tremor earthquake.

2. _____ A short-term earthquake may be too small to be felt.

3. _____ When the gas builds up, you have a long-term earthquake.

Drafting
Write a sentence to begin your explanation. Tell your main idea about the types of earthquakes you have researched.

_____
_____

Now write your explanatory essay on a separate piece of paper. Tell how each type of earthquake occurs. Use time-order or spatial words to make your explanation clear.

Revising and Proofreading
Here is part of Hector's explanatory essay. Add a time-order word or a spatial-order word in each blank to make the meaning clearer.

_____ magma pushes its way _____

to the surface, the rocks begin to break. _____

long-term earthquakes begin. Finally, magma flows

_____ the surface and breaks through.

Now revise and proofread your writing. Ask yourself:
▶ Did I discuss each type of earthquake and explain the difference?
▶ Did I include clear details that are easy to follow?
▶ Did I include time-order words or spatial order words?
▶ Did I correct all mistakes?

Name _____ Date _____ **LESSON Outline**

Shaping Earth's Surface

Use your textbook to help you fill in the blanks.

What is weathering?

1. The process that breaks down rock into small pieces is called _____.

2. Impacts, temperature changes, and ice expanding in cracks break down rock in the process of _____.

3. When rock's composition is broken down and changed, _____ occurs.

What is erosion?

4. Pieces of weathered rock are moved from place to place by _____.

5. When rock and soil on a slope become loose, gravity can move them downhill in a _____.

6. A large mass of flowing ice, called a _____, can erode the rock and soil beneath it.

What is deposition?

7. The process of _____ picks up eroded material and leaves it in a different place.

8. The running water of _____ erodes rock and soil and washes it downstream.

9. Slow-moving rivers can flow in gentle loops called _____.

Chapter 5 • Our Dynamic Earth
Reading and Writing

Use with **Lesson 5**
Shaping Earth's Surface

LESSON Outline

Name _____ Date _____

10. Waves wash away at the sides of a headland forming a(n) _____.

11. Waves can move sand along beaches or deposit it offshore to build strips of sand called _____.

How are shorelines changed?

12. During floods, rivers deposit sediment on _____ along their banks.

13. Deposits of sand along the shore can form _____ islands that protect the beaches behind them from storm waves.

14. Wind deposits sand along the shoreline in hills of sand called _____.

How can shorelines be protected?

15. People build walls called _____ along rivers to prevent floods.

16. To slow erosion _____ can be built in the water along the beach.

17. Fences and _____ protect dunes by preventing sand from blowing away.

Critical Thinking

18. How can people help stop erosion on a beach?

_____
_____
_____
_____

Name _____ Date _____

LESSON Vocabulary

Shaping Earth's Surface

Who am I? What am I?

Choose a word from the word box below that answers each question.

a. acid rain	d. erosion	g. meander
b. delta	e. floodplain	h. sediment
c. deposition	f. glacier	i. weathering

1. _____ I am a huge mass of ice. I erode the rock beneath me as I flow over it. Who am I?

2. _____ I am particles of rock. Moving water deposits me when it slows down. What am I?

3. _____ I wear away stone and metal surfaces when I fall from the sky. What am I?

4. _____ I am a flat area along a riverbank. Rivers deposit sediment on me when they flood. Who am I?

5. _____ I drop eroded sediment in a different place after I picked it up. What am I?

6. _____ I am a fan-shaped piece of land. I form from deposits at the mouths of rivers. Who am I?

7. _____ I break down rock into smaller pieces. What am I?

8. _____ I am a gentle loop in a slow-flowing river. What am I?

9. _____ I pick up and move pieces of weathered rock. What am I?

LESSON Cloze Activity

Shaping Earth's Surface

Fill in the blanks.

beaches	erosion	physical weathering
chemical weathering	glaciers	waves
deposition	landslides	weathering

Rocks are constantly being changed by the environment. Over long periods of time, the process of _____ breaks down rock. When _____ occurs, the rock simply breaks into smaller pieces. During _____, the rock weakens as the minerals in it are changed. After weathering, _____ removes the weathered rock. Gravity pulls loosened rock downhill in _____. Erosion and deposition shape shorelines. Ocean _____ and currents move sediment along shorelines. They erode sand from some _____ and deposit it on others.

As _____ move over the ground, they scoop out depressions and move the loose rock beneath them. Water, wind, and ice can drop eroded materials in a different place in a process called _____. This process also changes landforms.

Name _____ Date _____

Reading in Science

Wrestling with the Big Muddy

Read the Reading in Science feature in your textbook.

Problem and Solution

Identify the main problem described in the reading. Then identify the solution to the problem and the steps taken to reach it. Write the information in the correct boxes in the graphic organizer below.

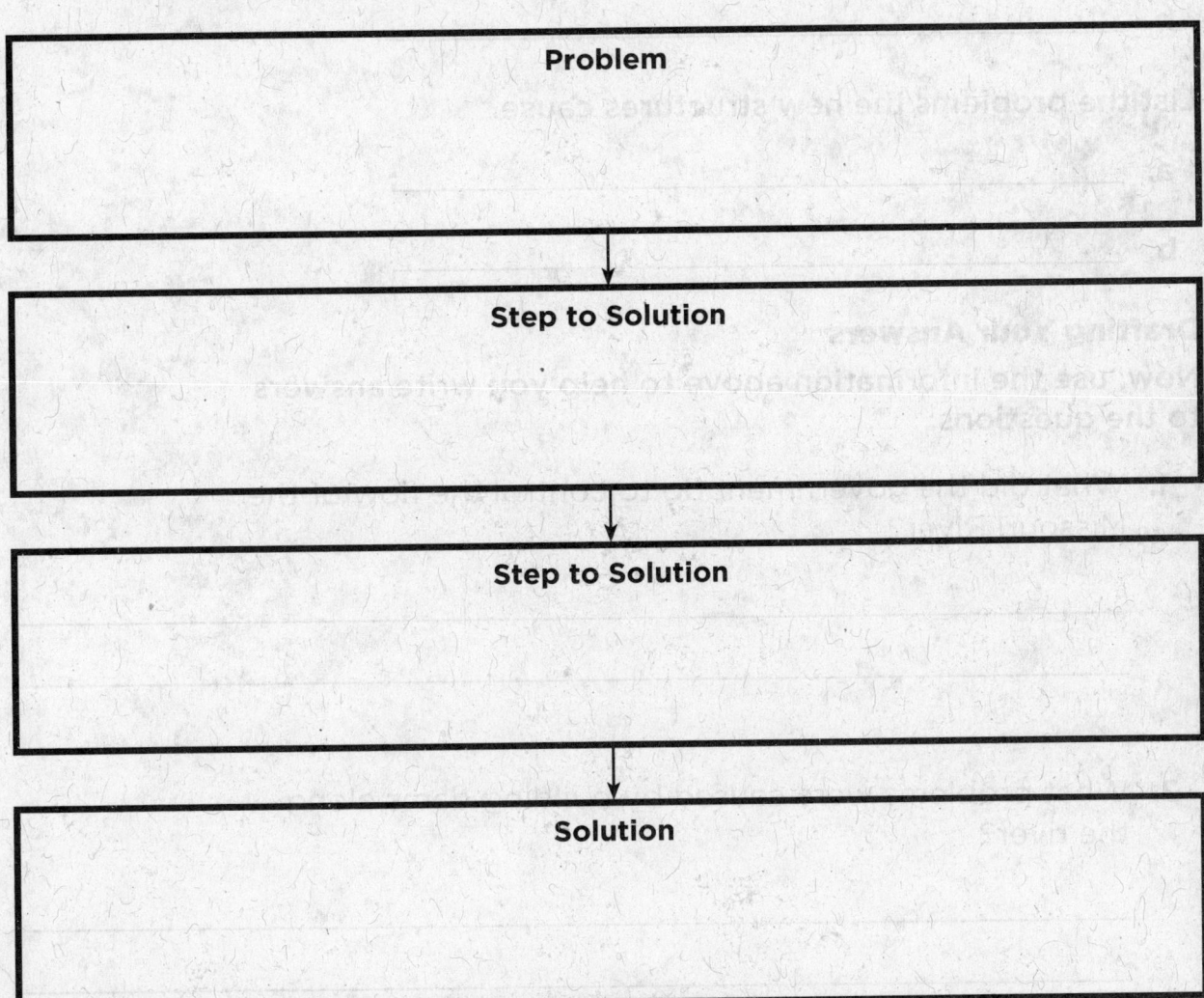

Chapter 5 • Our Dynamic Earth
Reading and Writing

Use with **Lesson 5**
Shaping Earth's Surface

123

Reading in Science Name _____ Date _____

Planning and Organizing

Read the "Write About It" questions carefully. Find the text within "Wrestling with the Big Muddy" that answers each question. Use the questions below to help organize your information.

List structures that now control the Missouri River.

a. _____

b. _____

List the problems the new structures cause.

a. _____

b. _____

Drafting Your Answers

Now, use the information above to help you write answers to the questions.

1. What did the government do to control the flow of the Missouri River?

   _____
   _____
   _____

2. What problems were caused by building dams along the river?

   _____
   _____
   _____

Our Dynamic Earth

Choose the letter of the best answer.

1. Which of these is a physical feature on Earth's surface?
   a. tsunami
   b. landform
   c. mantle
   d. hotspot

2. The crust and the top part of the mantle make up the
   a. atmosphere.
   b. hydrosphere.
   c. asthenosphere.
   d. lithosphere.

3. Earth's surface layer is the
   a. mantle.
   b. crust.
   c. biosphere.
   d. asthenosphere.

4. What layer of Earth's interior lies just below the crust?
   a. mantle
   b. inner core
   c. lithosphere
   d. outer core

5. The plate tectonics model states that Earth's crust is composed of
   a. one solid piece of rock.
   b. both liquid and frozen water.
   c. several huge slabs of rock that fit together.
   d. hot, melted rock.

6. What is a fault?
   a. energy that an earthquake produces
   b. the opening in a volcano
   c. a large crack in Earth's crust
   d. the boundary between two plates

7. Huge slabs of rock moving suddenly against each other in the Earth's crust create
   a. earthquakes.
   b. abyssal plains.
   c. volcanos.
   d. a trench stretch.

CHAPTER Vocabulary

Name _____ Date _____

8. A volcano is
   a. an opening in Earth's crust through which magma flows.
   b. any mountain near a plate boundary.
   c. a group of faults near a hot spot.
   d. movement at a fault.

9. The low area between mountains is called a
   a. plateau.
   b. trench.
   c. valley.
   d. landform.

10. The wide, flat are of the ocean floor is known as the
    a. mantle.
    b. abyssal plain.
    c. trench stretch.
    d. aquatic plateau.

11. What is the term used for melted rock that reaches the Earth's surface?
    a. lava
    b. mantle
    c. magma
    d. boundary rock

12. A device used to detect and measure earthquake waves is called a
    a. wavometer.
    b. richtometer.
    c. barometer.
    d. seismometer.

13. A mass of large flowing ice that can erode rock is called a(n)
    a. glacier.
    b. landslide.
    c. iceberg.
    d. delta.

14. Underwater earthquakes of a great magnitude can create
    a. continental divides.
    b. trenches.
    c. tsunamis.
    d. aquatic drift.

126 Chapter 5 • Our Dynamic Earth
Reading and Writing

Name _____ Date _____

CHAPTER Concept Map

Protecting Earth's Resources

Use your textbook to help you fill in the blanks.

Name of Resource	Soil	Energy	Water	Air
Different Types of Resource				N/A
Source of Resource		The Sun, wind, water, atoms, biomass, fossil fuels		
Uses for Resource			Animals and plants need water to live.	Animals breathe in oxygen from the air to stay alive.
Threats to Resource	Erosion	Overuse of non-renewable energy sources		
Ways to Protect Resource			Follow laws that prevent water pollution. Conserve water.	Reduce use of cars; decrease factory emissions.

Chapter 6 • Protecting Earth's Resources
Reading and Writing

LESSON Outline

Name _____ Date _____

Minerals and Rocks

Use your textbook to help you fill in the blanks.

What are minerals?

1. A solid natural substance underground made from nonliving materials is a(n) _____.

2. Minerals are made of one or more _____.

3. The color powder a mineral leaves when rubbed on a rough surface is its _____.

4. The way a mineral reflects light is its _____.

5. How well a mineral resists scratching is its _____.

6. Scientists use the _____ Scale to compare the hardness of minerals.

What are the shapes of a mineral?

7. The elements in minerals are in the form of _____, which are solids whose shapes form patterns.

8. Important minerals such as copper are found in _____, which are combinations of many minerals.

What is the rock cycle?

9. Over time, rocks change from one type to another in the _____.

10. Pressure can cement layers of weathered and eroded sediment into _____ rock.

Name _____ Date _____

LESSON Outline

11. When magma and lava cool and harden, they become _____ rock.

12. If they become buried deep beneath Earth's surface, sedimentary and igneous rocks can become _____ rock.

What are igneous and sedimentary rocks?

13. Igneous rocks that form inside Earth are called _____, and have _____ crystals.

14. Igneous rocks that form from lava that cools on Earth's surface are _____, and have _____ crystals.

What are metamorphic rocks?

15. When metamorphic rocks form, the shape and _____ of crystals can change, or the crystals can change position to form _____.

Critical Thinking

16. What are the different ways that rocks are produced, and what are the different properties of minerals?

_____
_____
_____
_____
_____

Chapter 6 • Protecting Earth's Resources
Reading and Writing

Use with **Lesson 1**
Minerals and Rocks

LESSON Vocabulary

Minerals and Rocks

Who am I? What am I?

Choose a word from the word box below that answers each question.

a. crystal	d. luster	g. rock cycle
b. hardness	e. metamorphic rock	h. sedimentary rock
c. igneous rock	f. mineral	

1. _____ I am the measure of how well a mineral resists scratching. What am I?

2. _____ I am a type of rock that forms when sedimentary and igneous rocks change under heat and pressure. Who am I?

3. _____ I am a solid natural material made from nonliving substances in the ground. What am I?

4. _____ I am a solid whose shape forms a pattern. What am I?

5. _____ I am the type of rock that forms from layers of sediment. Who am I?

6. _____ I am the way a mineral reflects light from its surface. What am I?

7. _____ I am the type of rock that forms from magma or lava that cools and hardens. Who am I?

8. _____ I am the change that occurs over time of one type of rock to another. What am I?

Minerals and Rocks

Fill in the blanks.

cleavage	lava	metamorphic
fractures	layers	minerals
igneous	luster	rock cycle

There are three categories of rocks. Rocks that form from cooled and hardened magma or _____ are _____ rocks. Rocks that form from _____ cemented together are sedimentary rocks. Heat and pressure deep inside Earth change igneous and sedimentary rocks into _____ rock. One rock can change into another type of rock in the _____ . All rocks are made from _____ that have many different properties. These properties include _____ , or the way the rock reflects light, and its color. A mineral is said to have _____ when it breaks along smooth surfaces. When it breaks along uneven surfaces, it _____ . The measure of how well a mineral resists scratching is its hardness.

LESSON Outline

Name _____ Date _____

Soil

Use your textbook to help you fill in the blanks.

What is soil?

1. Soil is a mixture of bits of _____ and once-living parts of plants and _____ .

2. The formation of soil starts with the _____ of rock.

3. Soil forms in layers that are called soil _____ .

4. The A horizon contains _____ which is made up of decayed organic materials.

5. The soil in the A horizon is also called _____ and is the soil in which most _____ grow.

6. The A horizon also contains the decayed organic materials, or _____ , that makes soil fertile.

7. The B horizon, called the _____ , has lots of fine rock particles but little humus.

8. The C horizon, which rests on _____ , is mostly large pieces of weathered rock.

How is soil used?

9. Soil in forests has a thin layer of _____ , and has little _____ .

Name _____ Date _____

LESSON Outline

10. Desert soil is sandy and does not hold much _____.

11. The soil of the prairies and other _____ in the central United States is rich in humus.

12. Grassland soil is good for _____.

13. Plants hold nutrients that return to the soil when the plants die and _____.

14. Chemicals that kill insects and weeds can cause soil to become _____.

How is soil conserved?

15. Farmers can replace humus and nutrients in soil with _____.

16. When farmers practice _____, they plant different crops on the same land in different years.

17. Farmers can conserve soil on hills when they use _____ plowing and _____.

Critical Thinking

18. What composes soil?

_____
_____
_____
_____
_____

Chapter 6 • Protecting Earth's Resources
Reading and Writing

Use with Lesson 2
Soil

LESSON Vocabulary

Name _____ Date _____

Soil

Use the words below to complete the sentences.

| bedrock | horizon | pollution | topsoil |
| conservation | humus | soil | |

1. The saving or protection of soil is _____.

2. The A horizon of soil, where most plants grow, is _____.

3. A mixture of particles of rock and bits of once-living parts of plants and animals is _____.

4. The part of soil made up of decayed materials is _____.

5. The adding of harmful materials to soil, air, or water is _____.

6. A layer of soil is a soil _____.

7. Large pieces of rock, on which the soil's C horizon rests, are called _____.

LESSON Cloze Activity

Soil

Fill in the blanks.

bedrock	large	pollution
desert	layers	subsoil
forest	plants	topsoil

Soil is a mixture of weathered rock and humus. It covers most of Earth's surface. Soil is divided into several _____ called soil horizons. There is unweathered _____ beneath the soil. On top of this layer is a C horizon with pieces of rock that are _____ in size. Above this is the B horizon or the _____. In this layer, there are small/fine rock particles and humus. The A horizon is the _____. It contains the most humus and is good for the growth of _____. There are mainly three types of soil in the United States: _____ soil, _____ soil, and grassland/prairie soil. Soil is a resource that can be spoiled by _____ from chemicals. It can also be eroded by flowing water and wind.

LESSON Outline

Name _____ Date _____

Fossils and Energy

Use your textbook to help you fill in the blanks.

What are fossils?

1. The remnants or traces of organisms from long ago that are preserved in soil or rock are _____.

2. Many fossils formed when organisms died and were covered with layers of _____.

3. Over millions of years, sediment covered and compressed dead plants to form soft or _____ coal.

4. Sometimes increased heat and pressure turned soft coal into harder _____ coal.

5. Heat and pressure on buried ocean plants and animals helped to form _____ and _____.

6. Coal, oil, and natural gas are _____.

How old are fossil and fossil fuels?

7. Scientists can tell how old a fossil is by testing the age of the _____ around it.

8. The law of superposition says that each layer of rock is _____ than the layer below it.

9. The comparison that tells whether one fossil is older than another fossil is _____.

Name _____ Date _____

LESSON Outline

How can wind, water, and the Sun provide energy?

10. Sources of energy other than fossil fuels are called _____ energy sources.

11. Running or falling water spins generators to make electricity in a(n) _____ plant.

12. Energy from the Sun is called _____ energy. This energy does not pollute.

What are other sources of alternative energy?

13. Changes in the centers of _____ can release heat that produces nuclear power.

14. Heat from deep inside the Earth is _____ energy that can produce electricity and provide hot water.

How can we conserve energy?

15. You use energy when you ride in a(n) _____ or use anything at home that runs on _____ .

16. When you do not waste energy, you _____ it.

Critical Thinking

17. How did ancient organisms become fossil fuels?

_____
_____
_____
_____
_____
_____

Chapter 6 • Protecting Earth's Resources
Reading and Writing

Use with **Lesson 3**
Fossils and Energy

Fossils and Energy

Fill in the blanks.

a. absolute age	e. fossil fuel
b. alternative energy	f. nonrenewable
c. era	g. relative age
d. fossils	h. renewable

1. Any source of energy other than fossil fuels is _____ .

2. The value that tells you whether a fossil is younger or older than another fossil is its _____ .

3. A resource that can be used up faster than it is made is _____ .

4. To find the _____ of a fossil, you must find the exact age of the rock that surrounds it.

5. The remnants or traces of ancient organisms that were preserved in soil or rock are _____ .

6. Resources that can be replaced faster than they are used are _____ .

7. A material formed from the decay of ancient organisms that is used to produce energy is a(n) _____ .

8. A unit of time that describes the age of Earth in millions of years is a(n) _____ .

Fossils and Energy

Fill in the blanks.

alternative energy	nonrenewable	Sun
coal	nuclear	water
geothermal	oil	
natural gas	pollution	

Fossil fuels are formed from the decay of ancient organisms. Examples of fossil fuels are _____, _____, and _____. These fossil fuels are _____ resources. We also use _____ sources, which are energy sources that are not fossil fuels. Renewable energy sources include wind, falling _____, and the _____. These forms of energy do not produce _____ that dirties the air and water. Another energy source is _____ energy, which comes from heat inside Earth. People also burn materials such as wood, a type of biomass. Changes in the nucleus of atoms release energy that runs _____ power plants. To save energy, people do things to conserve it.

Writing in Science

Name _____ Date _____

So You Want to Be a Fossil Hunter

Write About It
Select a fossil and write a description of it. Use sensory words and specific details in your description.

Getting Ideas

What fossil will you describe? Write its name in the center circle of the web below. Write details that describe the fossil in the outer circles. You can add circles to the web if you like.

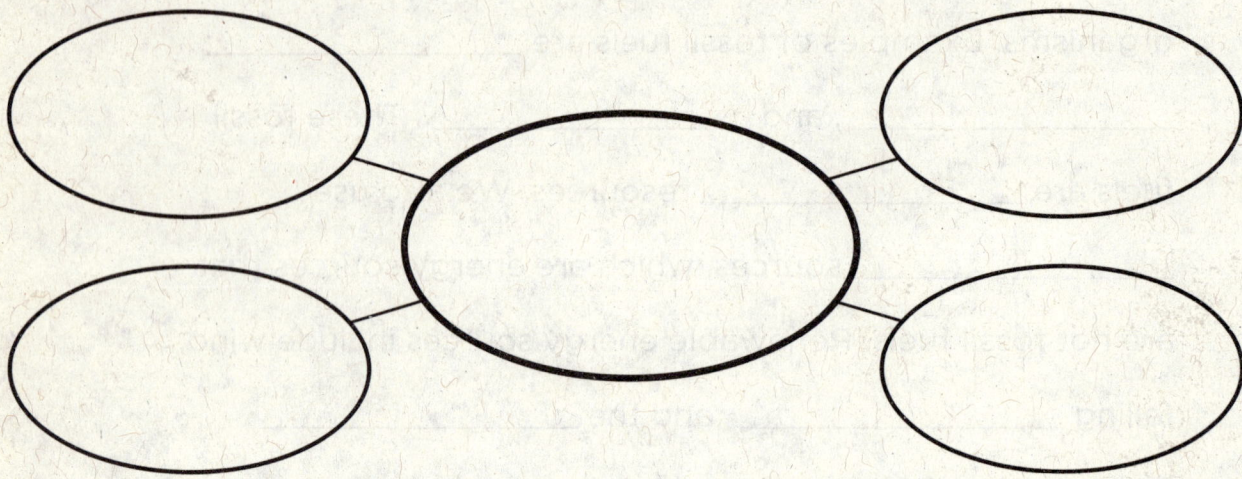

Planning and Organizing

Jorge wants to describe a fossil of a dinosaur footprint. Here are some sentences that he wrote. Write Yes if the sentence describes the fossil. Write No if it does not.

1. The huge footprint was $2\frac{1}{2}$ feet across. _____

2. It showed that the dinosaur had three long bony toes. _____

3. I got scared when I looked at the footprint. _____

Name _____ Date _____

Writing in Science

Drafting

Write a sentence to begin your description. Tell what fossil you will describe. Tell an important idea about this fossil.

_____

Now write your description. Use a separate piece of paper. Start with the sentence you just wrote. Then write your description. Use words that appeal to the senses. Use details that will help your readers picture the fossil.

Revising and Proofreading

Help Jorge improve his description. Add sensory words in the blanks. Choose a word from the box or pick your own.

| deep | gray | narrow | sharp | spiky |

The fossil footprint in the cold _____ stone reveals secrets of this creature that lived millions of years ago. The footprint had made a _____ impression in the earth. This suggested that the dinosaur was very big and heavy. It showed long _____ shapes at the end of the toes. Maybe this is where its _____ claws dug into the earth. The heel of the foot was _____ , not wide.

Now revise and proofread your writing. Ask yourself:

▶ Did I include enough details to help readers picture the fossil?
▶ Did I use sensory words to bring my description to life?
▶ Did I correct all mistakes?

Chapter 6 • Protecting Earth's Resources
Reading and Writing

Use with **Lesson 3**
Fossils and Energy

LESSON Outline Name _____ Date _____

Air and Water

Use your textbook to help you fill in the blanks.

What are sources of fresh water?

1. About 70 percent of Earth's surface is covered with water, with most of it in the _____ .

2. Salt enters much of Earth's water as rain and ocean waves wash over dirt and _____ .

3. Running water includes sources such as _____ and _____ .

4. Standing water includes sources such as _____ and _____ that fill holes in the ground.

5. Water beneath Earth's surface is _____ .

6. Groundwater collects underground in layers of rock or soil called _____ .

How do we use water?

7. Water can pick up substances that _____ or contaminate it as it falls through the sky or runs along the ground.

8. Wastes from mines and _____ can also pollute water.

How do we clean, conserve, and protect water?

9. The following steps clean drinking water in water treatment plants: coagulation, _____ , filtration, and _____ .

Name _____ Date _____

LESSON Outline

10. People can reduce their use of water through _____.

How do we use and pollute air?

11. Particles produced by cars and trucks can create a yellow haze in the air called _____.

12. Chemicals in old aerosol cans and old air conditioners can escape high into the atmosphere and destroy _____.

13. In some areas, pollution caused by smoke and gases from factories combines with rain to form _____ rain.

How do we protect our air?

14. Many pollutants are now banned or disposed of before they get into the air because of the _____ Act.

15. For example, vehicles have devices that limit the amount of pollutants that come out of _____ pipes.

Critical Thinking

16. Why are water and air important resources?

_____
_____
_____
_____
_____
_____
_____

Chapter 6 • Protecting Earth's Resources
Reading and Writing

Use with **Lesson 4**
Air and Water

LESSON Vocabulary

Name _____ Date _____

Air and Water

Match the correct letter with the description and fill in the blank with the correct answer.

> **a.** aquifer **d.** reservoir **g.** running water
> **b.** groundwater **e.** smog
> **c.** ozone hole **f.** oceans

1. _____ Salty water bodies containing most of Earth's water

2. _____ A thin spot in the layer of ozone

3. _____ A lake made by people that is used to store water

4. _____ An underground layer of rock or soil that can absorb water

5. _____ Water that is beneath Earth's surface

6. _____ A type of air pollution caused by particles from cars and factories

7. _____ The type of water that comes from rivers and streams

Name _____ Date _____

LESSON Cloze Activity

Air and Water

Fill in the blanks.

aquifers	groundwater	plants
food	oceans	reservoirs
fresh	oxygen	streams

Two of Earth's most important resources are water and air. Most of Earth's water is the salt water in _____ . However, people and most other living things need _____ water to survive. Most of the fresh water people use comes from running water, standing water, and _____ . We get running water from _____ and rivers. Standing water comes from lakes and _____ . We get groundwater from underground layers of rock and soil called _____ that absorb water. Living things also need gases, such as _____ , carbon dioxide, and nitrogen, from the atmosphere. Plants use carbon dioxide to make _____ . Bacteria in soil use nitrogen to make chemicals that _____ need. People can make water and air unusable when they release pollution.

Reading in Science

Getting the Salt Out

Read the following passage. Underline any sentence that identifies a problem. Circle any passages that mention possible solutions to those problems.

Why does California have water shortages when it is next to the Pacific Ocean? People cannot drink ocean water because of the salts in it.

The island of Santa Catalina lies off the coast of Southern California. It is completely surrounded by the Pacific Ocean. However, people use water from the ocean all the time—to water crops, to take showers, and even to drink. How can they use the salty ocean water? The water is converted from salty to fresh at the Santa Catalina desalination plant. Desalination means "to remove salts."

At the desalination plant, ocean water is taken from an ocean water well. Once it is moved into the plant, salt and other impurities are removed from the water. The fresh water that is produced can now be used by people.

The Santa Catalina plant is one of the few desalination plants in the United States that produces water for public use. Desalination is an expensive process that uses a lot of energy. Despite its cost, there are desalination plant projects all over the world, including places like Saudi Arabia and Japan. Desalination is generally used when a community has so little access to fresh water that they are willing to pay a high price to get it. Scientists continue to research cheaper and more effective ways to produce fresh water from ocean water.

Problem and Solution

▶ Identify the problem by looking for a conflict or an issue that needs to be resolved.

▶ Think about how the conflict or issue could be resolved.

Name _____ Date _____

Reading in Science

Problem and Solution

Fill in the problem-and-solution graphic organizer below. Use the underlined passages from the reading to help you.

```
┌─────────────────────────────────────────────────┐
│                    Problem                      │
│  People cannot drink or use ocean water because │
│  of the _____ it contains.             │
└─────────────────────────────────────────────────┘
                        ↓
┌─────────────────────────────────────────────────┐
│                Steps to Solution                │
│  Communities can build _____ that turn │
│  _____ into _____ .           │
└─────────────────────────────────────────────────┘
                        ↓
┌─────────────────────────────────────────────────┐
│                   Solution                      │
│  Fresh water from _____ can be used for│
│  _____ , for _____ , and for  │
│  _____ .                               │
└─────────────────────────────────────────────────┘
```

Write About It

Problem and Solution 1. What is in ocean water that prevents the people of Santa Catalina Island from drinking and using it directly from the ocean? 2. How do the people of Santa Catalina get fresh water?

Answer the following questions. Use clues from the graphic organizer to help you.

1. What is in ocean water that prevents the people of Santa Catalina Island from drinking and using it directly from the ocean? _____

2. How do the people of Santa Catalina get fresh water?

Chapter 6 • Protecting Earth's Resources
Reading and Writing

Use with **Lesson 4**
Air and Water

147

Protecting Earth's Resources

Choose the letter of the best answer.

1. A solid natural material in the ground made from nonliving substances is a(n)
   a. rock.
   b. aquifer.
   c. mineral.
   d. horizon.

2. Which type of rock is formed from layers of sediment?
   a. igneous
   b. magma
   c. sedimentary
   d. granite

3. Igneous rocks form from
   a. lava and magma.
   b. fossils.
   c. layers of sediment.
   d. humus.

4. Which of these causes the formation of metamorphic rock?
   a. an increase in water content
   b. very high temperatures
   c. the growth of crystals
   d. the splitting of atoms

5. A mixture of pieces of rock and bits of once-living parts of plants and animals is
   a. humus.
   b. rock.
   c. pollution.
   d. soil.

6. Which part of soil is formed from decayed materials?
   a. rock
   b. minerals
   c. humus
   d. topsoil

7. Soil in the A horizon is called
   a. topsoil.
   b. bedrock.
   c. humus.
   d. subsoil.

Choose the letter of the best answer.

8. Harmful chemicals added to air, water, or soil are
   a. luster.
   b. pollution.
   c. runoff.
   d. smog.

9. The remnants, or traces, of ancient organisms preserved in soil or rock are known as
   a. fossils.
   b. minerals.
   c. horizons.
   d. fuels.

10. Which of these is a nonrenewable energy resource?
    a. wind
    b. falling water
    c. oil
    d. biomass

11. Which of these is an alternative energy resource?
    a. coal
    b. natural gas
    c. the Sun
    d. oil

12. Which of these statements is true of a nonrenewable energy resource?
    a. Its supply will never run out.
    b. It is used up faster than it is made.
    c. It cannot be burned as fuel.
    d. It can be replaced faster than it is used.

13. An underground layer of rock or soil that can absorb water is a(n)
    a. aquifer.
    b. reservoir.
    c. soil horizon.
    d. well.

14. Which of these is a source of drinking water for people?
    a. acid rain
    b. ozone holes
    c. groundwater
    d. pools of magma

15. A yellow haze in the air caused by particles from cars and factories is
    a. oxygen.
    b. acid rain.
    c. carbon dioxide.
    d. smog.

UNIT Literature

Name _____ Date _____

Strong Storms

Read the Literature feature in your textbook.

Write About It

Response to Literature This article describes the damage caused by severe rainstorms in Los Angeles. Research the damage severe rainstorms can cause. Write a report about the effects of severe rainstorms. Include facts and details from this article and your own research.

_____
_____
_____
_____
_____
_____
_____
_____
_____
_____
_____
_____
_____
_____
_____

Name _____ Date _____

CHAPTER Concept Map

Weather Patterns

Complete the concept map about weather.

Weather

The average weather in a given region is called _____.

Weather is predicted by measuring _____ and making _____.

The variables that contribute to weather are air pressure, _____, cloud cover, _____, and wind speed.

Types of Cloud Cover

Name	Definition
_____ clouds	Clouds composed of ice crystals high in the sky.
Cumulus clouds	_____
_____ clouds	Layered clouds at low altitudes.
Fog	_____

Types of Precipitation

Name	Definition
_____	Liquid precipitation
Sleet	_____
_____	Water vapor that turns directly into ice crystals
Hail	Raindrops that freeze and then are moved up by wind.

Chapter 7 • Weather Patterns
Reading and Writing

The Atmosphere and Weather

Use your textbook to help you fill in the blanks.

How does the Sun warm Earth?

1. Sunlight strikes Earth with the most vertical angle at the _____.

2. An area near the _____ receives less energy from sunlight than an area of the same size near the _____.

What are the layers of the atmosphere?

3. When energy from the Sun hits the Earth, 50 percent is absorbed by _____, and 20 percent is absorbed or reflected by _____.

4. Particles of gas in the air pressing on Earth's surface create a force called _____.

What changes air pressure?

5. Atmospheric pressure decreases as altitude _____.

6. As humidity increases, air pressure _____.

What are global winds?

7. Winds that blow between 30°North and 30°South latitudes are called the _____.

8. Air pressure near the equator is _____ than air pressure near the poles, a fact that causes air to move from the _____ toward the _____.

Name _____ Date _____

LESSON Outline

9. Winds that blow south from the North Pole curve to the _____ because of the _____.

What are local winds?

10. During the day, the Sun heats land more quickly than it heats water, so a(n) _____ blows; during the night, water cools more slowly than land does, so a(n) _____ blows.

11. In the morning, valley breezes blow _____; in the afternoon, mountain breezes blow _____.

How do we measure air pressure and wind?

12. Air pressure is measured with a(n) _____; wind speed is measured with a(n) _____; wind direction is measured with a(n) _____.

Critical Thinking

13. How does Earth's shape affect global temperatures and wind patterns?

_____
_____
_____
_____
_____
_____
_____

LESSON Vocabulary

Name _____ Date _____

The Atmosphere and Weather

Who am I? What am I?

Choose a word from the word box below that answers each question.

a. air pressure	d. humidity	g. weather
b. atmosphere	e. insolation	
c. global wind	f. troposphere	

1. _____ I make the air feel dry or sticky. I am the amount of water vapor in the air. What am I?

2. _____ I am the layer of gases nearest Earth, where all weather takes place. What am I?

3. _____ Look out your window. I am the current condition of the atmosphere. What am I?

4. _____ I am the envelope of air surrounding Earth. What am I?

5. _____ You can count on me to blow steadily in predictable directions over very long distances. Who am I?

6. _____ I am the solar energy that reaches your planet. What am I?

7. _____ I am the weight of air pressing against you. What am I?

The Atmosphere and Weather

Fill in the blanks.

air pressure	equator	low air pressure
angle	high air pressure	poles
direct rays	less dense	troposphere

The condition of the atmosphere at any time and place is called weather. Weather occurs in the _____, the layer of the atmosphere closest to Earth. Global weather patterns are largely due to Earth's shape and the _____ at which sunlight strikes Earth in different places. The equator receives more _____ from the Sun, whereas the _____ receive very low angles of sunlight. Therefore, the temperature at the _____ is always higher than that at the poles.

The uneven heating of Earth causes differences in _____. Warm air is _____ and has a lower air pressure than does cold air. Air always flows from areas of _____ to areas of _____. Differences in air pressure cause global winds that blow in predictable directions over long distances.

LESSON Outline

Name _____ Date _____

Clouds and Precipitation

Use your textbook to help you fill in the blanks.

How do clouds form?

1. As water vapor rises, it becomes colder and _____ on particles of dust to form _____.

2. Clouds composed of ice crystals high in the sky are called _____.

3. Puffy clouds at middle altitudes are called _____.

4. Layered clouds at low altitudes are called _____.

5. A cloud close to the ground is called _____.

How does precipitation form?

6. Raindrops that fall through a layer of cold air can freeze to form _____.

7. At low temperatures, water vapor turns directly into solid crystals called _____.

8. Rainfall is measured with an instrument called a(n) _____.

What are air masses and fronts?

9. When a cold, dry, air mass meets a warm, moist, air mass, the cold air pushes the warm air _____, producing _____ weather.

Name _____ Date _____

LESSON Outline

10. On a weather map, blue triangles on a line represent a(n) _____ ; red half-circles on a line represent a(n) _____ .

11. Almost all weather fronts in North America are pushed from west to east by the _____ .

What are highs and lows?

12. Areas on a weather map that have the same air pressure are connected with lines called _____ .

13. Low pressure systems bring _____ weather; high pressure systems bring _____ weather.

What do weather maps tell you?

14. To make weather maps, meteorologists collect and analyze data such as _____ , _____ , and air pressure.

Critical Thinking

15. How and why do clouds form along a front?
_____
_____
_____
_____
_____
_____
_____

Chapter 7 • Weather Patterns
Reading and Writing

Use with **Lesson 2**
Clouds and Precipitation

Clouds and Precipitation

Choose a word from the word box below to finish the puzzle.

| cumulus | high | mass | precipitation |
| front | isobar | meteorologist | stratus |

Across

4. Cloud that is low and layered
6. Connects all places that have the same air pressure
7. Pressure system that brings cool, clear weather
8. Scientist who studies the atmosphere

Down

1. Puffy cloud
2. Meeting place between two air masses
3. Rain, sleet, hail, or snow
5. Large region of air that has a similar temperature and humidity throughout

Clouds and Precipitation

Fill in the blanks.

air mass	fog	snow	weather
clouds	front	stratus	
cumulus	sleet	warm front	

The formation of precipitation begins when water vapor condenses on dust particles, forming _____. Clouds form in different places and have different shapes—_____ forms close to the ground, layered _____ clouds form at low altitudes, and puffy _____ clouds form at middle altitudes. Cirrus clouds form at the highest altitudes. Water droplets grow larger until they become heavy enough to fall as rain, _____, or _____.

A large region of air that has similar temperatures and humidity throughout is called a(n) _____. As air masses move, they cause changes in the _____. A place where two different air masses meet is called a(n) _____. Warm air moving toward cold air is called a(n) _____. Cold air moving toward warm air is called a cold front.

LESSON Outline

Name _____ Date _____

Severe Storms

Use your textbook to help you fill in the blanks.

What are thunderstorms?

1. Updrafts of warm, moist air result in tall clouds called _____ .

2. During a thunderstorm, particles of rain and ice rub against one another as they rush upward and downward, creating _____ .

3. The discharge of static electricity in thunderclouds is seen as _____

4. Lightning suddenly raises the temperature of the air, causing the air to expand violently, and producing a sound known as _____ .

What are winter storms?

5. Winter storms often form when a(n) _____ air mass meets a(n) _____ air mass.

6. Blizzards are snowstorms with _____ mile per hour winds and _____ of a mile visibility.

What are tornados?

7. Tornados begin to form when warm air moves upward in a thunderhead, creating a(n) _____ area that draws more air inward and upward.

Name _____ Date _____

LESSON Outline

8. Air moving into the low pressure closure begins to spin, creating a(n) _____ , which becomes a(n) _____ when it reaches the ground.

What are hurricanes?

9. A tropical storm has _____ winds with a(n) _____ pressure area at its center.

10. When wind speeds reach more than 73 miles per hour, a tropical storm becomes a(n) _____ .

11. The three types of cyclones are _____ , _____ , and _____ .

How are storms tracked?

12. Weather stations around the world use instruments such as _____ , _____ , and rain gauges to measure local weather conditions.

13. Weather balloons collect data on _____ , _____ , and _____ at higher altitudes.

Critical Thinking

14. Explain why severe storms occur along fronts.

_____
_____
_____
_____
_____

Chapter 7 • Weather Patterns
Reading and Writing

Use with Lesson 3
Severe Storms

LESSON Vocabulary Name _____ Date _____

Severe Storms

Match the correct letter with the description.

a. blizzard	e. storm surge
b. cyclone	f. thunderstorm
c. ground blizzard	g. tornado
d. hurricane	h. whiteout

1. _____ blizzard that occurs when snow is no longer falling

2. _____ rainstorm with thunder and lightning

3. _____ snowstorm with winds of 35 miles per hour and visibility of a 1/4 mile

4. _____ tropical storm with wind speeds reaching more than 74 miles per hour

5. _____ zero visibility caused by heavy snowfall combined with strong updrafts and downdrafts

6. _____ bulge of water in the ocean, caused by hurricane winds

7. _____ any storm with a low pressure closure that causes the formation of a circular pattern of winds

8. _____ rotating funnel-shaped cloud with winds that blow up to 300 miles per hour

Name _____ Date _____

LESSON Cloze Activity

Severe Storms

Fill in the blanks.

center	lightning	thunderheads	tropical
front	polar	thunderstorm	tropical storm
hurricane	thunder	tornado	

Storms come in many forms. A severe storm that includes

_____ and _____ is called a(n)

_____ . Warm, moist air is pushed up by cold air

along a(n) _____ , and _____ form.

Sometimes a thunderstorm can turn into a twister, or

_____ . A thunderstorm can also become

a(n) _____ , with rotating winds and a low

pressure area at its _____ . Such a storm can

turn into a(n) _____ .

Winter storms often form when a continental

_____ air mass meets a maritime

_____ air mass. A winter storm can drop many

forms of precipitation.

Chapter 7 • Weather Patterns
Reading and Writing

Use with **Lesson 3
Severe Storms**

Writing in Science

Living Through a Mudslide

Write About It
Write a personal narrative about a storm, mudslide, or other severe weather condition that you have experienced. Use a clear sequence of events to tell what happened and what you did.

Getting Ideas

Choose a severe weather condition you have experienced. Write its name in the center circle. Then put on your thinking cap. Write words and details that tell about this weather condition in the outer circles.

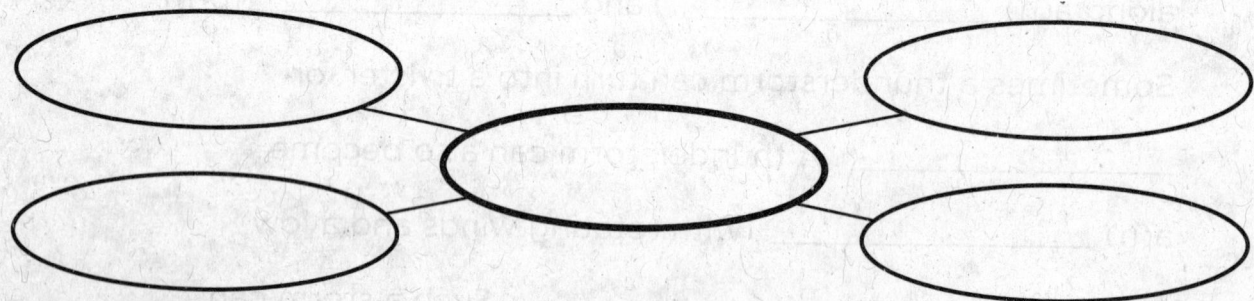

Planning and Organizing

Here are some sentences that Kevin wrote to tell about his experience during a hurricane. Number the sentences from 1-4, by 1 being the sentence that comes first.

_____ Next, the winds picked up, knocked over garbage cans, and tossed the trash like balls in the air.

_____ First, the sky grew dark as a wall of clouds marched in.

_____ Then, the waves built, growing higher and higher, until they crashed over the railings along Shore Road.

_____ Finally, Mom and Dad moved us all to the shelter before the full force of the storm hit.

Name _____ Date _____

Writing in Science

Drafting
Write a sentence to begin your personal narrative. Introduce yourself by using the pronoun "I." Name the weather condition and tell how it made you feel.

_____

_____

_____

Now write your personal narrative. Use a separate piece of paper. Begin with the sentence you wrote above. Tell about the events in time order. Use time-order words to make the sequence easy to follow.

Revising and Proofreading
Here is part of Kevin's personal narrative. He made five mistakes in grammar. Find the mistakes and correct them. Cross out the error. Write the correction above it.

> It started out as a beautiful day in late September. The sun was shining bright and the temperature were mild. My friends and I think it would be a great day for a bike ride along Shore Road. Was we ever wrong! My sister heard the announcement first and calls me into her room.

Now revise and proofread your writing. Ask yourself:
- ▶ Did I use the pronoun "I" to identify myself?
- ▶ Did I tell the events in sequence?
- ▶ Did I correct all mistakes in grammar, spelling, punctuation, and capitalization?

Chapter 7 • Weather Patterns
Reading and Writing

Use with **Lesson 3**
Severe Storms

Climate

Use your textbook to help you fill in the blanks.

What is climate?

1. Two variables that are important in determining climate are _____ and _____ .

2. The global variable that has the strongest effect on climate is _____ .

3. Areas along the equator are located in the _____ zone.

4. A way to categorize an area's climate is to describe the _____ that grow there.

5. Many scientists are concerned that the global climate is _____ .

6. Radiated heat from Earth's surface is _____ by a layer of greenhouse gases. Some of the heat then radiates back and warms Earth.

7. Greenhouse gases include _____ , _____ , and _____ .

8. Burning _____ increases the amount of greenhouse gases in the atmosphere, a factor in _____ .

Name _____ Date _____

LESSON Outline

What affects climate?

9. The temperature of an inland city is usually _____ in summer and _____ in winter than the temperature of a coastal city.

10. At a given latitude, the higher the altitude, the _____ the climate.

11. The climate on the _____ side of a mountain is wetter and cooler than the climate on the _____ side.

What is El Niño?

12. A cold current along the coast of Peru causes air pressure to be _____ in the eastern Pacific than it is in the western Pacific.

13. El Niño brings _____ to the coasts of North and South America; La Niña brings _____ to these coastal areas.

Critical Thinking

14. Location A is near the equator on the windward side of a mountain. Location B is at 30°N latitude on the east side of the Atlantic Ocean. Describe the climate in each location. Explain your answers.

_____
_____
_____
_____
_____

Chapter 7 • Weather Patterns
Reading and Writing

Use with Lesson 4
Climate   167

LESSON Vocabulary

Name _____ Date _____

Climate

Choose a word from the word box below to complete the puzzle.

| climate | Gulfstream | polar | tropical |
| ENSO | La Niña | temperate | windward |

Across

2. Climate zone located along the equator
3. Average weather of a place
5. Climate zone located at the North and South poles
7. Wetter side of a mountain
8. Comings and goings of El Niño

Down

1. Ocean current that warms Europe
4. Climate with warm summers and cold winters
6. The dryer weather that occurs when the current along the Peruvian coast sinks

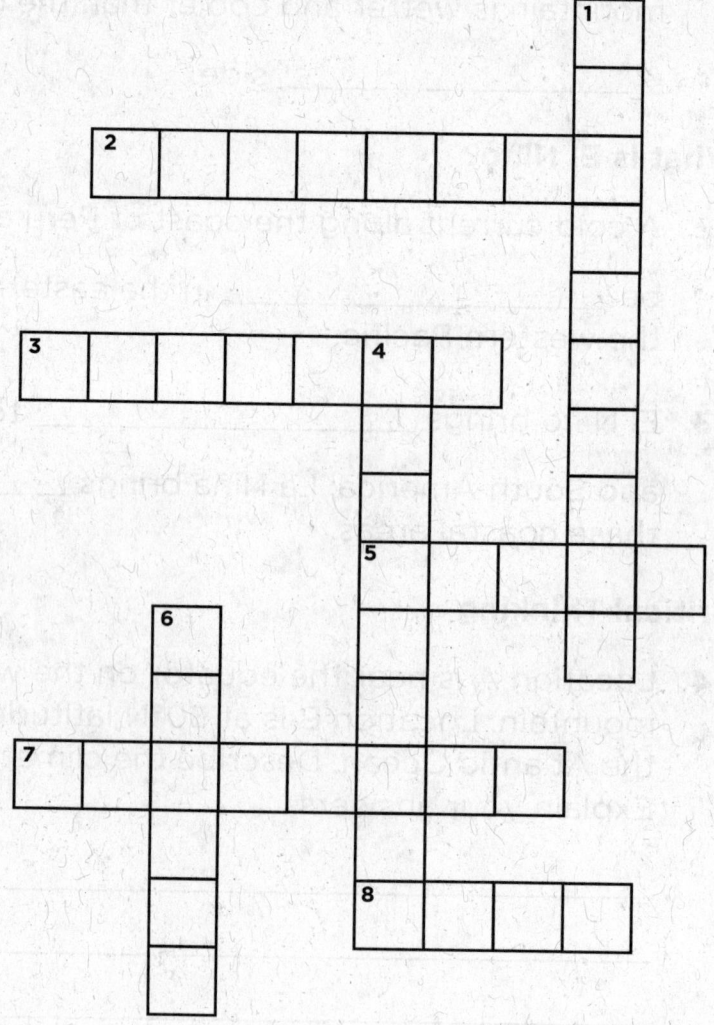

Name _____ Date _____

LESSON Cloze Activity

Climate

Fill in the blanks.

altitude	precipitation
body of water	temperature
latitude	temperate
ocean currents	tropical

The type of weather that exists in a place over the long term is its climate. The two most important variables that determine climate are _____ and _____. It is possible to predict the climate of an area if you know its _____.

Areas near the equator have _____ climates and the highest temperatures. They also have heavy precipitation during at least part of the year. Areas near the poles have polar climates. Areas between the tropical and polar zones have _____ climates. Other factors that affect climate are distance from a(n) _____, _____, and _____. All of these factors can give you a general idea of the climate of an area.

Chapter 7 • Weather Patterns
Reading and Writing

Use with Lesson 4
Climate 169

Reading in Science

Museum Mail Call

Read the following letters from the Reading in Science passage in your textbook. Underline the sentences or phrases that describe the features of each area.

June 13

Dear Museum Scientists,

Hola! (That's "hello" in Spanish) It's the dry season here in Palmdale right now and it's *muy caliente*—very hot! We haven't had rain in weeks.

It's usually hot and dry here from May to November. We don't have a lot of water, so it has to be piped in from other areas. Restaurants only serve water to people who ask for it.

Some people plant cactuses and shrubs around their homes. I planted jalapeño peppers with *mi hermana*, my sister. We water the plants in the evening. That way the hot sun won't dry up all of the water.

Carlos

June 23

Dear Museum Scientists,

The *gio mua*, or monsoons, have brought wet weather to our land. Everything here is soaked! Our monsoon season lasts from May to October. Many inches of rain can fall during heavy storms. But the storms only last for about an hour each day. It's very hot, so we don't mind getting wet.

Our farm is near the Mekong River. Water floods our rice fields and helps the rice grow. It's hard work walking through the swampy ground. We carry the rice with *quang ganh*. These are baskets that we balance at the end of a pole.

People here are used to a lot of water. We build our houses on stilts so the water won't get in. Some years, there is more water than we expect!

Vang

Name _____ Date _____

Reading in Science

Compare and Contrast

Fill in the Venn diagram below with the facts that you underlined in each of the letters on the previous page.

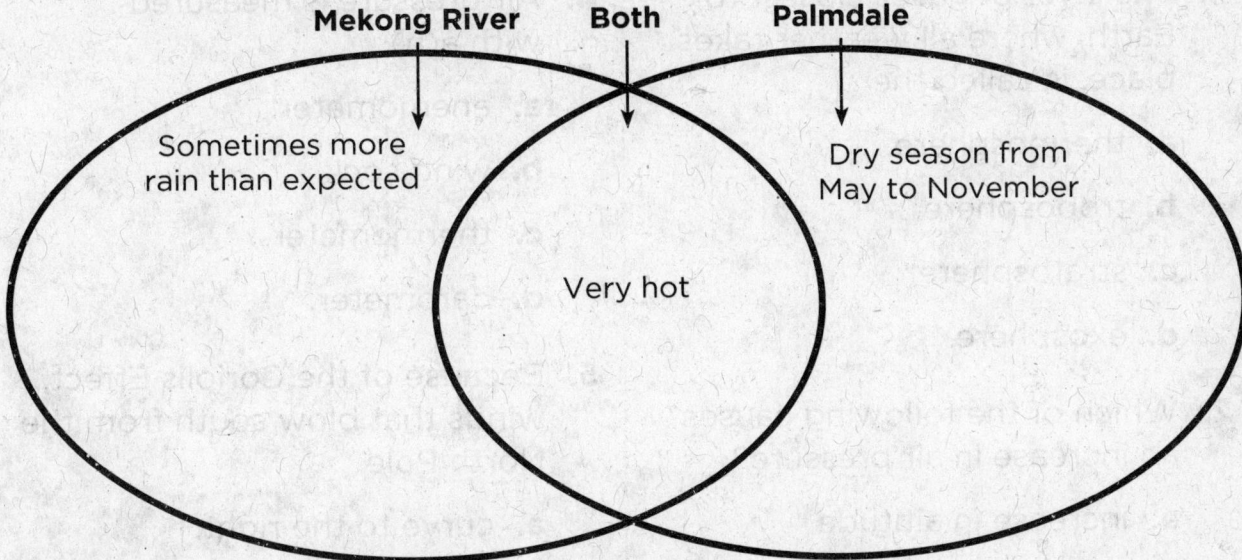

Mekong River — Sometimes more rain than expected

Both — Very hot

Palmdale — Dry season from May to November

Write About It

Compare and Contrast How does the weather in Palmdale compare with the weather near the Mekong River? What activity do both Carlos and Vang do?

Compare and Contrast

Answer the following questions, using the information you have about both Palmdale and the Mekong River.

1. How does the weather in Palmdale compare with the weather near the Mekong River?

   _____

   _____

2. What activity do both Carlos and Vang do?

   _____

   _____

Chapter 7 • Weather Patterns
Reading and Writing

Use with **Lesson 4**
Climate   171

CHAPTER Vocabulary

Weather Patterns

Choose the letter of the best answer.

1. The layer of gases closest to Earth, where all weather takes place, is called the
   a. thermosphere.
   b. troposphere.
   c. stratosphere.
   d. exosphere.

2. Which of the following causes an increase in air pressure?
   a. increase in altitude
   b. increase in volume
   c. increase in humidity
   d. decrease in temperature

3. Global winds occur because
   a. air pressure near the poles is lower than air pressure near the equator.
   b. sunlight heats areas near the equator more than it heats areas near the poles.
   c. sunlight warms the air over land faster than it warms the air over water.
   d. sunlight warms the air over mountains faster than it warms the air in valleys.

4. Air pressure is measured with a(n)
   a. anemometer.
   b. wind sock.
   c. thermometer.
   d. barometer.

5. Because of the Coriolis Effect, winds that blow south from the North Pole
   a. curve to the right.
   b. curve to the left.
   c. speed up.
   d. slow down.

6. A cloud close to the ground is called
   a. a cumulus cloud.
   b. a stratus cloud.
   c. fog.
   d. a cirrus cloud.

Name _____ Date _____

CHAPTER
Vocabulary

7. Which of the following best describes how snow forms?

   a. Water vapor freezes directly into a solid.

   b. Water droplets freeze and then fall as precipitation.

   c. Water droplets collide with bits of ice and freeze.

   d. Water droplets fall through a layer of cold air close to the ground.

8. An air mass that forms over northern Canada will be

   a. cold and humid.

   b. cold and dry.

   c. warm and humid.

   d. warm and dry.

9. Which of the following best describes how the weather will change when a cold front moves into an area?

   a. The weather will become drier.

   b. The weather will become clear and cool.

   c. The weather will become stormy, but when the front passes, the weather will become cool and dry.

   d. The weather will become stormy and warmer.

10. Which of the following is a cyclone?

    a. thunderstorm

    b. blizzard

    c. ice storm

    d. hurricane

11. When do storm surges occur?

    a. during a blizzard

    b. during a hurricane

    c. during a thunderstorm

    d. during a tornado

12. A storm that has an eye and rotating winds that reach 74 miles per hour is called a

    a. tropical storm.

    b. cyclone.

    c. tornado.

    d. hurricane.

13. A sudden discharge of static electricity during a thunderstorm is called

    a. thunder.

    b. lightning.

    c. a low pressure closure.

    d. a downdraft.

Chapter 7 • Weather Patterns
Reading and Writing

CHAPTER Concept Map

Name _____ Date _____

The Universe

Complete the concept map with information you learned about the universe.

The _____ is a huge space that holds energy and matter.

↓

Most of the matter is in groupings of stars, dust, and gas called _____ , which can be spiral, _____ , or irregular.

↓

The spiral _____ that you live in is called the _____ .

↓

This galaxy includes the _____ , which has the Sun at its center.

↓

The solar system has eight _____ that orbit the Sun. These include Mercury, _____ , Earth, Mars, _____ , Saturn, Uranus, and Neptune. Many of these have natural satellites called _____ .

↓

The solar system also has rocky asteroids and icy _____ that orbit the Sun.

---

174 Chapter 8 • The Universe
Reading and Writing

Name _____ Date _____

LESSON Outline

Earth and Sun

Use your textbook to help you fill in the blanks.

What is gravity?

1. The force of attraction between any two objects is _____ .

2. The strength of gravity increases as the _____ of objects increases and decreases as the distance between objects increases.

3. The Sun's gravitational pull on Earth is _____ than its pull on a planet much farther away, such as Neptune.

4. A path that one object takes as it moves around another object is its _____ .

5. Earth and the other planets move in orbits around the _____ .

6. Moving objects have the tendency to keep moving in a straight line; this is called _____ .

7. Because of the effects of gravity and inertia, Earth moves in a nearly circular orbit shaped like a(n) _____ .

What causes the seasons?

8. Every year, Earth makes one complete trip, or _____ , around the Sun.

9. As Earth revolves around the Sun, sunlight strikes different parts of Earth at different _____ .

Chapter 8 • The Universe
Reading and Writing

Use with **Lesson 1** Earth and Sun

LESSON Outline Name _____ Date _____

10. Sunlight strikes Earth differently at different times of year because Earth's axis is _____.

11. The changes in the angle of sunlight on Earth's surface cause the _____.

12. When the Northern Hemisphere is tilted away from the Sun, the season there is _____.

13. When it is winter in the Northern Hemisphere, it is _____ in the Southern Hemisphere.

14. The heat energy of sunlight is more concentrated in the summer because the hemisphere having summer is tilted _____ the Sun.

What causes day and night?

15. As Earth revolves around the Sun, it also _____ on its axis.

16. At any time, half of Earth faces the Sun and has _____ , while the other half faces away and has _____.

Critical Thinking

17. If Earth's axis were not tilted, could any area have both a hot summer and a cold winter?
   _____
   _____
   _____

176 Chapter 8 • The Universe
Reading and Writing

Use with **Lesson 1**
Earth and Sun

Name _____ Date _____

LESSON Vocabulary

Earth and Sun

Who am I? What am I?

Choose a word from the word box below that answers each question.

a. ellipse	**d.** orbit
b. gravity	**e.** revolution
c. inertia	**f.** rotation

1. _____ I am the tendency of a moving object to keep moving in a straight line. What am I?

2. _____ I am the spinning of Earth around its axis. I cause day and night. Who am I?

3. _____ I am the path that one object, such as a planet, takes as it moves around another object. What am I?

4. _____ I am one complete trip around the Sun. For Earth, one of me is a year. Who am I?

5. _____ I am the force of attraction, or pull, between two objects. I increase as the mass of the objects increases. What am I?

6. _____ I am the shape of a planet's orbit. Who am I?

Chapter 8 • The Universe
Reading and Writing

Use with **Lesson 1**
Earth and Sun

LESSON Cloze Activity

Earth and Sun

Fill in the blanks.

axis	night	summer
concentrated	revolution	winter
day	seasons	

Earth moves in two main ways. Each year, it makes one _____ around the Sun. At the same time, Earth also spins on its _____ . As it spins, half of Earth faces the Sun and has _____ , while the other half faces away from the Sun and has _____ . Earth's tilt on its axis causes _____ .

The hemisphere of Earth tilted toward the Sun has _____ . Temperatures are warmer at this time of year because the Sun's heat strikes at a direct angle and is _____ . The hemisphere tilted away from the Sun has _____ . The seasons in the Northern Hemisphere and Southern Hemisphere are always opposite.

Name _____ Date _____

LESSON Outline

Earth and Moon

Use your textbook to help you fill in the blanks.

How does the Moon appear?

1. Although the Moon has no water, vast plains called _____, a Latin word meaning "seas," cover large parts of its surface.

2. Rocks striking the Moon over billions of years have formed many _____.

3. The Moon shines with light that comes from the _____ and reflects off the Moon's surface.

4. The appearance and shape of the Moon as you see it from Earth is called a(n) _____.

What causes eclipses?

5. A darkening or hiding of the Sun, a planet, or a moon by another object in space is a(n) _____.

6. A solar eclipse happens when the _____ casts a shadow on part of Earth, and people there see the Moon move across the face of the Sun.

7. A solar eclipse occurs only during the _____ phase.

8. A lunar eclipse happens when the Moon moves into and is hidden by the shadow of _____.

Chapter 8 • The Universe
Reading and Writing

Use with **Lesson 2**
Earth and Moon

LESSON Outline

Name _____ Date _____

9. During an eclipse, the area where the Sun is completely blocked is the _____, and the area where light is not completely blocked is the _____.

What causes the tides?

10. The rise and fall of the ocean's surface because of the gravity of the Moon and Sun is the _____.

11. A bulge of water occurs on the side of Earth facing the _____ and on the opposite side of the planet.

12. There are high tides at the bulges of water and _____ halfway between the bulges.

13. When the Sun and Moon align at full moon and pull on Earth together, the higher high tides and lower low tides are called _____.

14. When the Sun and Moon pull at a right angle and their pulls partly cancel each other, more moderate tides called _____ occur.

Critical Thinking

15. What would be different on Earth if the Moon did not exist?

_____
_____
_____
_____

Chapter 8 • The Universe
Reading and Writing

Use with **Lesson 2**
Earth and Moon

Name _____ Date _____

LESSON Vocabulary

Earth and Moon

Use the words below to help you complete the sentences.

| lunar eclipse | phase | solar eclipse |
| maria | rill | tide |

1. A groove in the Moon's surface is often called a _____.

2. The periodic rise and fall of the ocean's surface is the _____.

3. The appearance and shape of the Moon as you see it from Earth is called a _____.

4. When the Moon moves into Earth's shadow, a _____ occurs.

5. A vast plain on the Moon's surface is a _____.

6. When the Moon passes directly between the Sun and Earth, a _____ happens.

Chapter 8 • The Universe
Reading and Writing

Use with **Lesson 2**
Earth and Moon

Earth and Moon

Fill in the blanks.

low	month	rotates	tides
lunar eclipse	phase	solar eclipse	

The Sun lights half of the Moon at all times, but people on Earth see different amounts of the Moon's lit half at different times of the _____. The shape of the Moon as you see it from Earth at a particular time is called its _____. The Moon sometimes passes directly between the Sun and Earth, causing a _____. When Earth's shadow falls on the Moon, a _____ occurs.

The gravity of the Moon and the Sun pull on the surface of Earth's oceans, forming bulges that we know as _____. As Earth _____, the tides move around the planet. Most places have two high tides and two _____ tides during a single day.

The Moon is a rocky body with no atmosphere.

Name _____ Date _____

Writing in Science

What Would Happen if Gravity Went Away?

Read the Writing in Science feature in your textbook.

Write About It

Explanatory Writing You know that gravity keeps everything on Earth from floating off into space. Look at the picture on page 438 of your textbook. Explain what would change if gravity suddenly stopped working.

Planning and Organizing

Explanatory writing requires you to organize your ideas in chronological or time order. When Luis planned to make a mobile to represent the solar system, he needed to list the steps in sequence. Here are some steps that he wrote, number them from 1 to 5 with 1 being the first step.

1. _____ Next, cut out the circles. Punch a hole at the top.

2. _____ Then, thread the string through the hole in each circle. Attach it to a coat hanger. Finally, paste a cutout of the Sun onto the coat hanger.

3. _____ First, look at the sizes of the planets in comparison to each other.

4. _____ After that, use string to represent how far each planet is from the Sun.

5. _____ Then, use a compass to draw circles on cardboard to represent each planet. Make sure each circle represents the relative size of each planet. Color each planet and write its name.

Chapter 8 • The Universe
Reading and Writing

Use with **Lesson 2**
Earth and Moon   183

Writing in Science

Name _____ Date _____

Now write the first draft of your composition. Begin with a paragraph that establishes your topic and briefly describes the important ideas. Then describe the events that occur in chronological order. End with a short summary of the events and how they relate to your topic.

_____
_____
_____
_____
_____
_____
_____
_____
_____
_____
_____
_____
_____
_____

Now revise and proofread your instructions. Ask yourself:

- ▶ Have I explained the topic and described the important ideas?
- ▶ Have I described the events in time order?
- ▶ Have I provided clear descriptions of the events?
- ▶ Have I corrected all grammar errors?
- ▶ Have I corrected all errors in spelling, punctuation, and capitalization?

Name _____ Date _____

LESSON Outline

The Solar System

Use your textbook to help you fill in the blanks.

How do we observe objects in space?

1. An optical telescope uses _____ or mirrors to make distant objects seem larger and nearer.

2. The orbiting Hubble Space Telescope "sees" objects more clearly than Earth-based telescopes because Earth's _____ does not change Hubble's view.

3. Radio telescopes are giant dishes on Earth's surface that gather _____ waves from objects in space.

What are planets?

4. The solar system includes eight _____ that orbit the Sun.

5. The planet closest to the Sun is _____, and the planet farthest away from the Sun is _____.

6. Mercury, Venus, Earth, and Mars are terrestrial planets with surfaces made of _____.

7. Jupiter, Saturn, Uranus, and Neptune have surfaces made of _____.

How do the planets compare?

8. The most noticeable feature about _____ is its large set of rings that are made of ice and rock.

Chapter 8 • The Universe
Reading and Writing

Use with Lesson 3
The Solar System

LESSON Outline Name _____ Date _____

9. Venus has an atmosphere made mostly of _____, which holds in heat and gives this planet the hottest surface in the solar system.

10. The solar system's highest mountains and largest canyon system are on the surface of the planet _____.

How do the moons compare?

11. A natural object that orbits a planet is a _____.

12. The solar system's moons are natural _____, but artificial satellites orbit Earth to gather weather data and help people communicate.

13. Objects from space can strike moons or planets to create bowl-shaped holes called _____.

What are asteroids, comets, and meteors?

14. Most of the solar system's _____ orbit the Sun in a belt between Mars and Jupiter.

15. A comet is a mixture of _____, dust, and rock that moves around the Sun.

How do we explore the solar system?

16. The only place in the solar system that humans have visited is Earth's _____.

Critical Thinking

17. Why is it not possible to land a spacecraft on Jupiter or Saturn?

_____

_____

Chapter 8 • The Universe
Reading and Writing

Use with Lesson 3
The Solar System

The Solar System

Match the correct letter with the description and fill in the crossword puzzle.

- a. asteroid
- b. comet
- c. meteor
- d. moon
- e. planet
- f. satellite
- g. telescope

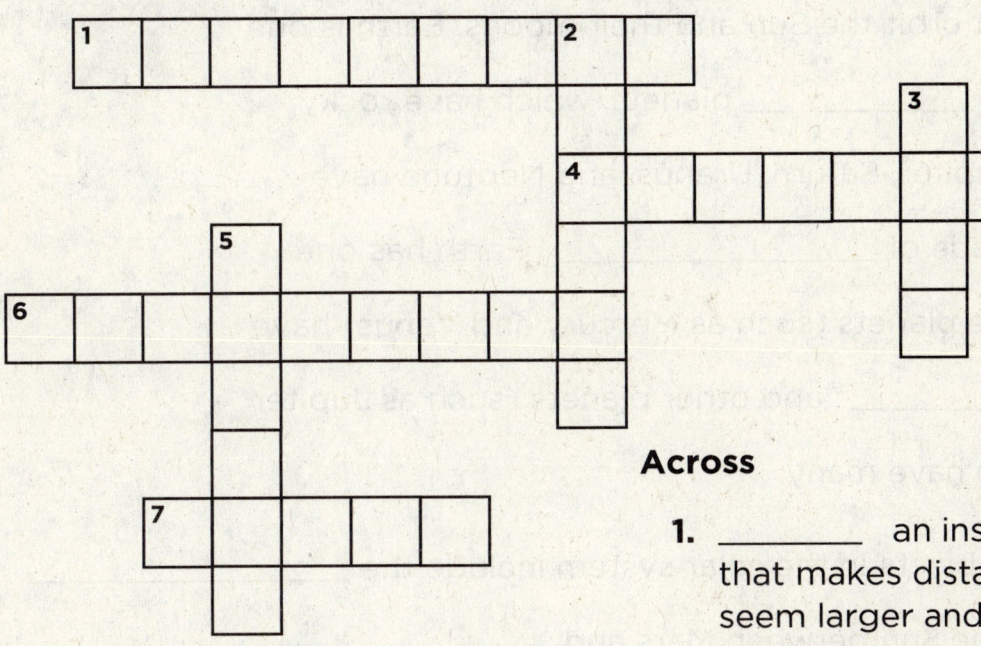

Across

1. _____ an instrument that makes distant objects seem larger and nearer

4. _____ a rock that revolves around the Sun in a belt between Mars and Jupiter

6. _____ an object in space that orbits another object

7. _____ an object made of ice, dust, and rock that orbits the Sun

Down

2. _____ a large object that orbits a star but does not give off its own light

3. _____ a natural object that orbits a planet

5. _____ a meteoroid that enters Earth's atmosphere

LESSON Cloze Activity

The Solar System

Fill in the blanks.

| asteroids | gases | none | telescopes |
| comets | Jupiter | tail | terrestrial |

The major objects of the solar system are eight planets that orbit the Sun and their moons. Earth is one of the _____ planets, which have rocky surfaces. Jupiter, Saturn, Uranus, and Neptune have surfaces made of _____ . Earth has one moon, some planets (such as Mercury and Venus) have _____ , and other planets (such as Jupiter and Saturn) have many.

Other objects in the solar system include the _____ that orbit the Sun between Mars and _____ . Balls of ice, dust, and rock in elongated elliptical orbits around the Sun are _____ . When far away from the Sun, comets remain frozen, but they form a glowing _____ of gas and dust as they get close to the Sun. Astronomers study the solar system with many types of _____ .

Voyager Discoveries

Read the following passage.

In 1977, NASA launched the Voyager Interstellar Mission to explore Jupiter, Saturn, Uranus, Neptune, and their moons. The trip had to be very precisely planned. Speeds and distances had to be accurately calculated. The two *Voyager* spacecraft had to be close enough to each planet to collect data and to get a pull from that planet's gravity in order to be propelled toward their next destination. At the same time, the spacecraft had to be far enough away from the planets that they would not go into orbit around them. All of NASA's careful planning worked. The *Voyager* Mission has provided scientists with new and closer looks at our farthest neighbors.

Voyager Spacecraft Travel

Jupiter–1979:
Images show Jupiter's rings. Volcanic activity is observed on Io, one of Jupiter's moons.

Saturn–1980-91:
Scientists get a close look at Saturn's rings. They contain structures that look like spokes, or braids. Scientists observed that Titan, one of Saturn's moons, has a thin atmosphere and active, geyser-like landforms.

Uranus–1986:
Scientists discover the dark rings around Uranus. They also see ten new moons, bringing Uranus's total to 15 moons. *Voyager* sends back detailed images and data on the planet, its moons, and dark rings.

Neptune–1989:
Large storms are seen on the planet. One of these storms is Neptune's Great Dark Spot. Neptune was originally thought to be too cold to support this kind of weather.

Reading in Science

Name _____ Date _____

After observing these planets, the *Voyager* spacecraft keep traveling. They are the first human-made objects to go beyond the heliosphere. The heliosphere is the region of space reached by the energy of our Sun. It extends far beyond the most distant planets in the Solar System.

Write About It
Cause and Effect

▶ Look for the reason why something happens to find a cause.

▶ An effect is what happens as a result of a cause.

1. What would cause the *Voyager* spacecraft to be propelled toward their next destination?

   _____
   _____
   _____
   _____
   _____

2. What was an effect of the *Voyager* mission?

   _____
   _____
   _____
   _____
   _____

Chapter 8 • The Universe
Reading and Writing

Use with Lesson 3
The Solar System

Name _____ Date _____

LESSON Outline

Stars and the Universe

Use your textbook to help you fill in the blanks.

How do stars form?

1. Stars form from a huge cloud of gases and dust called a(n) _____.

2. When the cloud contracts and powerful reactions start to turn hydrogen atoms into helium atoms to produce energy, a(n) _____ forms.

3. A _____ is a small very dense star that shines with cool white light.

What happens to larger stars?

4. A star that begins life with much more hydrogen than a medium-size star such as our Sun ends its life as an exploding star called a(n) _____.

5. Very massive stars end their lives as _____, which are objects with gravity so powerful that even light cannot escape from them.

6. Stars are classified by their size, _____, and temperature.

7. The Sun is a medium-size _____ star with a surface temperature of about 6,000°C.

8. By studying the motion of distant stars, scientists have discovered about 160 _____ outside our solar system.

Chapter 8 • The Universe
Reading and Writing

Use with **Lesson 4**
Stars and the Universe   191

What are constellations?

9. Patterns of stars in the sky are _____.

10. Most stars are so far from Earth that astronomers use huge measuring units, such as the _____, to describe the distance.

What are star systems?

11. Huge, far-off families of stars that look like hazy patches of faint light in the night sky are _____.

12. There are spiral galaxies, _____ galaxies, and irregular galaxies.

13. When two stars are near each other and rotate around each other, they form a(n) _____ star.

How did the universe form?

14. The theory that the universe started from a single point and then exploded outward is the _____ theory.

15. According to this theory, the universe continues to _____.

Critical Thinking

16. Will the Sun always shine?

_____

_____

_____

Name _____ Date _____

LESSON Vocabulary

Stars and the Universe

Match the correct letter with the description.

a. big bang theory	d. galaxies	g. star
b. black hole	e. light-year	h. supernova
c. constellation	f. nebula	i. white dwarf

1. An exploding star is a _____ .

2. Huge, very far-off families of stars are _____ .

3. An object in space that produces its own energy, including heat and light, is a _____ .

4. The idea that the universe began with a big bang and has been expanding since that time is the _____ .

5. The distance that light travels in one year is a _____ .

6. An object that is so dense and has so much gravity that not even light can escape it is a _____ .

7. A huge cloud of gases from which stars form is a _____ .

8. A group of stars that forms a pattern is a _____ .

9. A small, very dense star is a _____ .

Chapter 8 • The Universe
Reading and Writing

Use with Lesson 4
Stars and the Universe

LESSON Cloze Activity

Stars and the Universe

Fill in the blanks.

| elliptical | helium | spiral | white dwarf |
| galaxies | nebulas | Sun | 10 billion |

Scientists use the big bang theory to explain how the universe began and why it is expanding. The universe contains many families of stars called _____. Those shaped like pinwheels are _____ galaxies. There are also _____ galaxies and irregular galaxies. The star closest to Earth, after the _____, is Proxima Centauri.

Like living things, stars have life cycles. Stars are born from clouds of gas called _____. When gravity causes nebulas to contract enough, temperature rises and reactions that change hydrogen into _____ start. When the helium is also gone, the star shrinks and cools to become a _____. The life cycle of a medium-size star, such as our Sun, is about _____ years. Our Sun is about 5 billion years old.

The Universe

Choose the letter of the best answer.

1. Gravity is the
   a. measure of mass.
   b. force of attraction between objects.
   c. size of an object.
   d. long distance between stars.

2. What is an orbit?
   a. the speed of a planet moving around the Sun
   b. the order of planets in distance from the Sun
   c. the path a planet takes as it moves around the Sun
   d. the tilt of Earth on its axis

3. Inertia is the tendency of a moving object to
   a. keep moving after it hits something.
   b. keep moving in a straight line.
   c. keep moving faster and faster.
   d. rise upward against gravity.

4. What is Earth's revolution?
   a. its spinning motion on its axis
   b. its gravitational pull on the Moon
   c. its changing of seasons
   d. its movement in orbit around the Sun

5. Which of these is a lunar phase?
   a. penumbra
   b. new moon
   c. lunar eclipse
   d. neap moon

6. What happens during new moon when the Moon passes directly between the Sun and Earth?
   a. a solar eclipse
   b. a quarter moon
   c. a lunar eclipse
   d. a new season

7. A lunar eclipse happens only during
   a. new moon.
   b. crescent moon.
   c. quarter moon.
   d. full moon.

CHAPTER Vocabulary

Name _____ Date _____

8. What causes tides?
   a. earthquakes beneath the ocean
   b. the gravity of the Moon and Sun
   c. Earth's inertia in space
   d. high winds on the ocean's surface

9. A natural object that orbits a planet is a(n)
   a. asteroid.
   b. comet.
   c. moon.
   d. star.

10. In the solar system, most asteroids are
    a. beyond Neptune.
    b. orbiting Saturn.
    c. between Mars and Jupiter.
    d. next to the Sun.

11. When a meteor lands on the surface of Earth, it is called a(n)
    a. asteroid.
    b. meteorite.
    c. comet.
    d. satellite.

12. What does a telescope do?
    a. makes objects in space appear larger and nearer
    b. brings objects closer to Earth
    c. makes Earth seem brighter
    d. allows us to see black holes

13. Stars form from a cloud of gas called a
    a. galaxy.
    b. nebula.
    c. universe.
    d. neutron star.

14. What object is so dense and has such strong gravity that no light can escape it?
    a. black hole
    b. neutron star
    c. white dwarf
    d. red giant

15. What is the name of the theory that explains the way the universe began?
    a. The Gravitational Microlensing Theory
    b. The Stellar Life Cycle Theory
    c. The Big Bang Theory
    d. The Expanding Universe Theory

Chapter 8 • The Universe
Reading and Writing

Name _____ Date _____

UNIT Literature

Green and Clean: Plants as Pollution Control

Read the Literature feature in your textbook.

Write About It

Response to Literature This article describes how plants are used to help clean polluted soil. Research additional information about cleaning up waste. Write a report about the cleaning process. Include facts and details from this article and from your research.

_____
_____
_____
_____
_____
_____
_____
_____
_____
_____
_____
_____
_____
_____
_____
_____
_____

Comparing Kinds of Matter

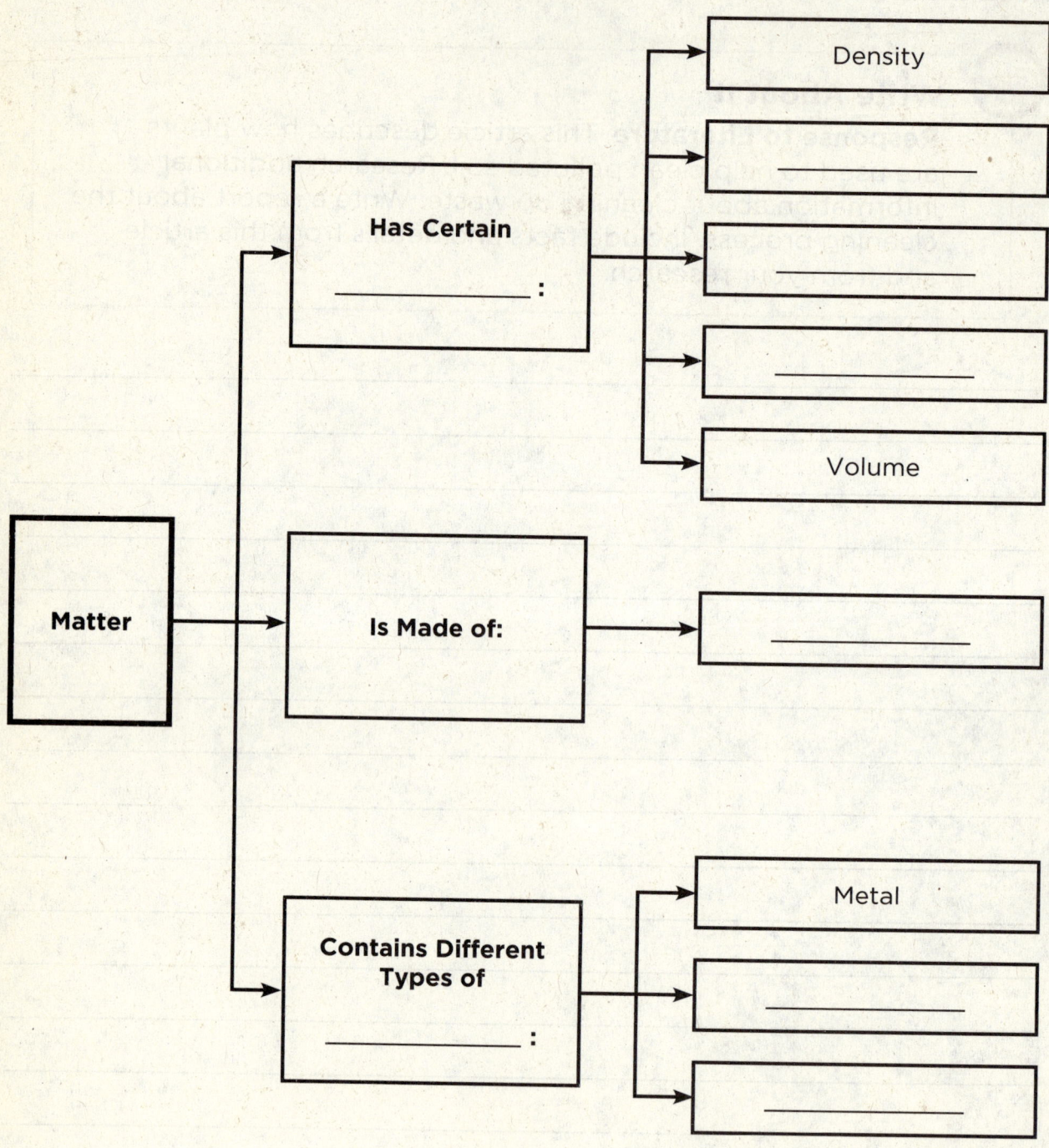

Name _____ Date _____

LESSON Outline

Properties of Matter

Use your textbook to help you fill in the blanks.

How can you describe matter?

1. The amount of matter in an object is its _____.

2. The mass of an object is measured in _____ or kilograms.

3. A measure of how strongly gravity pulls on an object is the object's _____.

4. The greater the _____ of an object, the greater its weight.

5. Weight is measured in _____.

6. The amount of space an object takes up is its _____.

7. To measure liquid volume in _____, scientists use tools such as beakers or graduated cylinders.

8. The volume of solids is measured in _____.

9. Anything that has mass and volume is _____.

What is density?

10. The amount of mass for each milliliter of a substance is that substance's _____.

11. To calculate density, divide an object's _____ by its _____.

Chapter 9 • Comparing Kinds of Matter
Reading and Writing

Use with **Lesson 1**
Properties of Matter

LESSON Outline

Name _____ Date _____

12. Buoyancy depends on _____, which depends on mass and volume.

13. Changing the mass or volume of an object changes its density and _____.

14. If an object covers a large enough area of the water's surface, it can float on the water because of the _____ of water particles.

What forms can matter have?

15. Matter can exist as a solid, a(n) _____, or a gas.

16. A solid has a definite _____ and volume.

17. A liquid has a definite _____, but it takes the shape of the container holding it.

18. A gas does not have a definite volume or a definite _____.

Critical Thinking

19. How can matter be described?

_____
_____
_____
_____
_____
_____
_____

Properties of Matter

Fill in the crossword puzzle from the clues below.

| buoyancy | mass | Newton | volume |
| density | matter | surface tension | weight |

Across

4. The amount of space that matter takes up
5. The metric unit used to measure weight
6. Anything that has mass and volume
7. The amount of mass for each milliliter of a substance

Down

1. The property of water that helps certain objects float
2. An object's resistance to sinking
3. How strongly gravity pulls on an object
6. The amount of matter in an object

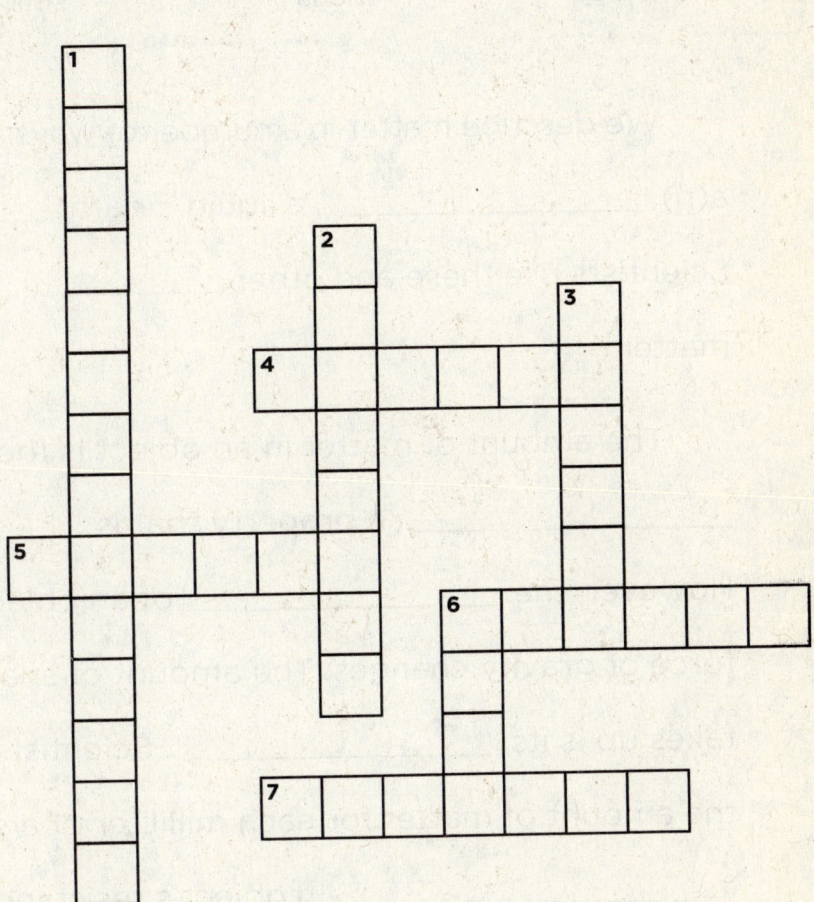

LESSON Cloze Activity

Name _____ Date _____

Properties of Matter

Fill in the blanks.

buoyancy	float	properties	solid
constant	gas	push	volume
density	mass	sink	weight

We describe matter in a number of ways. Matter can exist as a(n) _____, a liquid, or a(n) _____. Scientists use these and other _____ to identify matter.

The amount of matter in an object is the object's _____, a property that is _____. However, the _____ of an object changes as the force of gravity changes. The amount of space that an object takes up is its _____. Scientists also measure the amount of matter for each milliliter of a substance, or its _____. An object's resistance to sinking is _____. When an object is placed on a fluid, the object and the fluid _____ against each other. If the fluid is denser, the object will _____. If the object is denser, the object will _____. Matter is anything that has mass and volume.

202 Chapter 9 • Comparing Kinds of Matter
Reading and Writing

Use with Lesson 1
Properties of Matter

Name _____ Date _____

LESSON Outline

Elements

Use your textbook to help you fill in the blanks.

What is matter made of?

1. A substance that cannot be broken down chemically into simpler substances is a(n) _____.

2. One important property of an element is its _____ at room temperature.

3. Another important property of an element is the way that it _____ with other elements.

4. Today we know that a(n) _____ is the smallest unit of an element that has that element's properties.

What are atoms and molecules made of?

5. The center of an atom is its _____.

6. An atom's nucleus contains particles called protons that have a positive charge and particles called _____ that have no charge.

7. Negatively charged particles called _____ move around the nucleus.

8. Because an atom has the same number of _____ and electrons, the atom has no overall charge.

9. The number of protons in an atom is that atom's _____.

Chapter 9 • Comparing Kinds of Matter
Reading and Writing

Use with Lesson 2 **203**
Elements

LESSON Outline

Name _____ Date _____

10. An atom's protons and neutrons have about the same mass, which is one _____ unit, or amu.

11. If you add up the mass of all the protons and neutrons in an atom, you get the atom's _____.

12. Two or more atoms joined into a single particle form a(n) _____.

13. Molecules have properties that are different than the _____ that form them.

How are elements grouped?

14. Dmitri Mendeleev created the _____ of elements.

15. The table's columns group elements according to their _____.

What are the most common elements?

16. In space, the most common elements are _____ and helium.

17. On Earth, elements such as _____ and any of these: oxygen, silicon, aluminum, nitrogen, iron, and calcium are among the most common.

Critical Thinking

18. What is matter made of?

_____
_____
_____

Name _____ Date _____

LESSON Vocabulary

Elements

Read each clue. Write the answer in the blanks using the words below.

| atom | element | molecule | nucleus |
| electron | metal | neutron | proton |

1. The smallest unit of an element that retains that element's properties is a(n) _____.

2. The particle in an atom that has a negative charge is a(n) _____.

3. A substance that chemical reactions cannot break down into something simpler is a(n) _____.

4. An element that has properties such as shine, conductivity, and flexibility is a(n) _____.

5. Two or more atoms that are joined into one particle are a(n) _____.

6. In the nucleus of an atom, a particle that has no electrical charge is a(n) _____.

7. The center of an atom is its _____.

8. In the nucleus of an atom, a particle that has a positive electrical charge is a(n) _____.

Chapter 9 • Comparing Kinds of Matter
Reading and Writing

Use with **Lesson 2** Elements

Elements

Fill in the blanks.

atoms	metal	nonmetal	properties
electrons	metalloid	nucleus	temperature
elements	neutrons	periodic table	

Every substance on Earth is made of one or more _____. Dmitri Mendeleev created the _____ in the 1860s. It groups elements according to their _____. One important property of an element is its state at room _____. Another is the way that it combines or mixes chemically with other elements. A third property is the element's classification as a(n) _____, a(n) _____, or a(n) _____.

Each element is composed of tiny particles called _____, the smallest units that retain the element's properties. All atoms have the same parts. The center of an atom is its _____. The nucleus contains protons and _____. Atoms also contain _____, which move around the nucleus. Protons and neutrons have a much larger mass than electrons do.

Element Discovery

When Mendeleev shuffled his element cards to create the periodic table in 1869, he suspected he wasn't playing with a full deck. Many of the elements had already been discovered, but he believed others would come later.

1766 Hydrogen—The most abundant atom in nature is discovered by Henry Cavendish. In 1766, Cavendish is experimenting with materials in his lab when he isolates a gas that is flammable. He realizes that this gas might be a new element and calls it flammable air. The element later gets its name from the Greek words meaning "water forming," when another scientist discovers that water is made of hydrogen and oxygen.

1772–74 Oxygen—Scientists Joseph Priestley and Carl Wilhelm Scheele independently discover that when they heat certain compounds, a new kind of "air" or gas is given off. The new gas makes substances burn five times faster than ordinary air. The new gas is named oxygen from the Greek words meaning "acid former." That's because when oxygen combines with other elements, the compounds are usually acidic.

1868–1895 Helium—Joseph Lockyer discovers helium in 1868 by studying the Sun's spectrum with a spectroscope during a solar eclipse. He finds color lines that no element at the time was known to produce. He infers the lines must be due to a new element found only in the Sun. The element is named helium, after Helios, the Greek god of the Sun. In 1895, helium is finally found on Earth in uranium minerals.

1940 Plutonium—Scientists in Berkeley, California, create a new element by bombarding uranium with particles of deuterium, a special form of hydrogen. They name the element after the recently discovered planetary body Pluto.

1952 Einsteinium—A team of scientists find this element while studying the radioactive debris created when a hydrogen bomb explodes. They name it in honor of scientist Albert Einstein. Only a small amount of einsteinium has ever been produced, and it exists only for a fraction of a second before it transforms itself into other elements.

The periodic table isn't finished. Elements are still being added to it. In the past 75 years, 26 new elements have been added to the table. That's about one element every three years! If you found a new element, what would you name it?

Write About It
Classify

1. Which elements were discovered as gases?
2. Which elements have names that describe their properties? How are the other elements named?

___

208 Chapter 9 • Comparing Kinds of Matter
Reading and Writing

Use with **Lesson 2**
Elements

Metals, Nonmetals, and Metalloids

Use your textbook to help you fill in the blanks.

What are metals?

1. Metals share certain properties, such as _____ surfaces.

2. Metals conduct _____ and _____ well.

3. Metals are also easy to shape because they have _____ .

4. The property of _____ allows a metal to be pulled into thin wires.

5. Almost all metals occur naturally in the _____ state, but they vary in _____ .

6. When left out of doors, many metals will _____ as they combine with nonmetals around them.

How do we use metals?

7. Metals such as steel are useful because they are both _____ and flexible.

8. Reactive metals such as cadmium and nickel are used to make electricity in _____ .

What elements are nonmetals and metalloids?

9. Nonmetals are not good _____ of electricity.

10. Instead of bending, nonmetals usually _____ or _____ .

LESSON Outline Name _____ Date _____

11. Elements with properties that are between metals and nonmetals are _____.

12. Solid metalloids look like metals, but they do not have _____ surfaces.

13. Because they do not bend well, metalloids are not _____ or ductile.

14. Metalloids are called _____ because they do not conduct electricity as well as metals but conduct it better than nonmetals.

How do we use nonmetals and metalloids?

15. Nonmetals are excellent _____ of electricity and heat.

16. Semiconductor metalloids such as _____ are used to make computer chips.

Critical Thinking

17. Describe the properties of metals, nonmetals, and metalloids.

_____
_____
_____
_____
_____
_____
_____
_____

Name _____ Date _____

LESSON Vocabulary

Metals, Nonmetals, and Metalloids

Who am I? What am I?

Choose a word from the word box below that answers each question.

a. corrosion	d. metal	g. nonmetal
b. ductility	e. metalloid	h. semiconductor
c. malleability	f. noble gas	

1. _____ I am a shiny solid that conducts electricity very well. What am I?

2. _____ I am very particular. I am an element that does not like to mix with others. What am I?

3. _____ Look for me in the middle of columns in the periodic table. I am located between the metals and the nonmetals. Who am I?

4. _____ I am a property of metals. Because of me, people can make copper into thin wires. What am I?

5. _____ I am a poor conductor of electricity. Try to bend or flatten me, and I will break or crumble instead. Who am I?

6. _____ I happen when metals are left outdoors and combine with nonmetals. I create rust in iron. Who am I?

7. _____ I am the property that lets you bend and shape a metal. What am I?

8. _____ I am a metalloid used in computer chips. I conduct electricity better than a nonmetal, but not as well as a metal. Who am I?

Chapter 9 • Comparing Kinds of Matter
Reading and Writing

Use with **Lesson 3**
Metals, Nonmetals, and Metalloids

Metals, Nonmetals, and Metalloids

Fill in the blanks.

break	electricity	metals
ductile	insulators	nonmetals
dull	malleable	opposite

Scientists classify an element as a metal, a nonmetal, or a metalloid on the basis of the element's properties. Most _____ can be polished until their surfaces are shiny. They conduct _____ and heat well. When bent and pulled, metals are both _____ and _____ .

Nonmetals have properties that are the _____ of those of metals. The surface of a nonmetal is _____ rather than shiny. Nonmetals are good _____ rather than conductors. Instead of bending, nonmetals _____ or crumble. Metalloids have some properties like those of metals and some that are more like those of _____ . Metalloids are semiconductors—materials that conduct electricity better than nonmetals do, but not as well as metals.

Name _____ Date _____

CHAPTER Vocabulary

Comparing Kinds of Matter

Choose the letter of the best answer.

1. A material that cannot be broken down into simpler chemical substances is a(n)
    a. element.
    b. metal.
    c. chemical.
    d. molecule.

2. What is the smallest particle of an element?
    a. molecule
    b. proton
    c. atom
    d. metalloid

3. The positively charged particles in an atom are called
    a. neutrons.
    b. electrons.
    c. protons.
    d. molecules.

4. Which particles share the nucleus of an atom with the protons?
    a. neutrons
    b. protons
    c. elements
    d. electrons

5. Which particles in an atom are negatively charged?
    a. protons
    b. neutrons
    c. molecules
    d. electrons

6. Two or more atoms can join to form a(n)
    a. element.
    b. neutron.
    c. molecule.
    d. superatom.

7. The amount of matter in an object is its
    a. weight.
    b. mass.
    c. volume.
    d. density.

Chapter 9 • Comparing Kinds of Matter
Reading and Writing

CHAPTER Vocabulary

Choose the letter of the best answer.

8. The pull of gravity on an object determines that object's
   a. mass.
   b. volume.
   c. weight.
   d. electrical charge.

9. The amount of space being taken up by matter is known as its
   a. volume.
   b. weight.
   c. mass.
   d. density.

10. Anything that has mass and volume is
    a. metallic.
    b. matter.
    c. gaseous.
    d. atomic.

11. The amount of mass for each milliliter of a substance determines the substance's
    a. weight.
    b. buoyancy.
    c. density.
    d. volume.

12. An object's resistance to sinking is called
    a. weight.
    b. buoyancy.
    c. volume.
    d. surface tension.

13. The property that allows matter to be bent, flattened, or hammered without breaking is
    a. malleability.
    b. surface tension.
    c. ductility.
    d. buoyancy.

14. What happens to a metal that is left exposed to the environment and combines chemically with a nonmetal?
    a. It shrinks.
    b. It becomes a metalloid.
    c. It corrodes.
    d. It becomes a nonmetal.

Name _____ Date _____

CHAPTER Concept Map

Physical and Chemical Changes

Use your textbook to help you fill in the blanks.

A physical change may involve a change in shape, size, or _____ of matter. The three states of matter are _____ , _____ , and _____ .

Name of Process	Speed of Process	Initial Phase	Final Phase
	Slow	Liquid	Gas
Boiling		Liquid	
		Solid	Gas
		Solid	Liquid
	Slow/Fast	Gas	Liquid

Chapter 10 • Physical and Chemical Changes
Reading and Writing

LESSON Outline

Name _____ Date _____

Changes of State

Use your textbook to help you fill in the blanks.

How can matter change state?

1. Altering the form or organization of an object without changing the type of matter within it is called a(n) _____ .

2. The three states of matter are _____ , liquid, and _____ .

3. The state of matter of an object is a(n) _____ property.

4. The average vibration of molecules in an object is measured by _____ .

5. When a solid gains heat energy, its molecules begin vibrating too quickly to stay together, so the solid becomes a(n) _____ .

6. When gases lose heat, they _____ into liquids.

7. A liquid loses heat and _____ into a solid.

8. When a solid changes directly into a gas, it _____ .

9. Most liquids become _____ when they change to a solid.

Name _____ Date _____

LESSON Outline

When does matter change states?

10. When a substance melts or boils, it absorbs _____.

11. The temperature at which a substance changes from a solid to a liquid is its _____.

12. The temperature at which a substance changes from a liquid to a gas is its _____.

13. The temperature at which a substance changes from a liquid to a solid is its _____.

14. Nonmetals are weakly attracted to one another, so they have _____ melting and boiling points.

15. The slow change from a liquid to a gas at temperatures below the boiling point is called _____.

What are expansion and contraction?

16. An increase in an object's volume when it is heated is called _____; a decrease in its volume when it is cooled is called _____.

Critical Thinking

17. How does water change when heat is added or removed?

_____
_____
_____
_____

Chapter 10 • Physical and Chemical Changes
Reading and Writing

Use with **Lesson 1**
Changes of State

LESSON Vocabulary

Changes of State

Choose words from the word box below to finish the crossword puzzle.

boiling   freezing   sublimation
contraction   melting

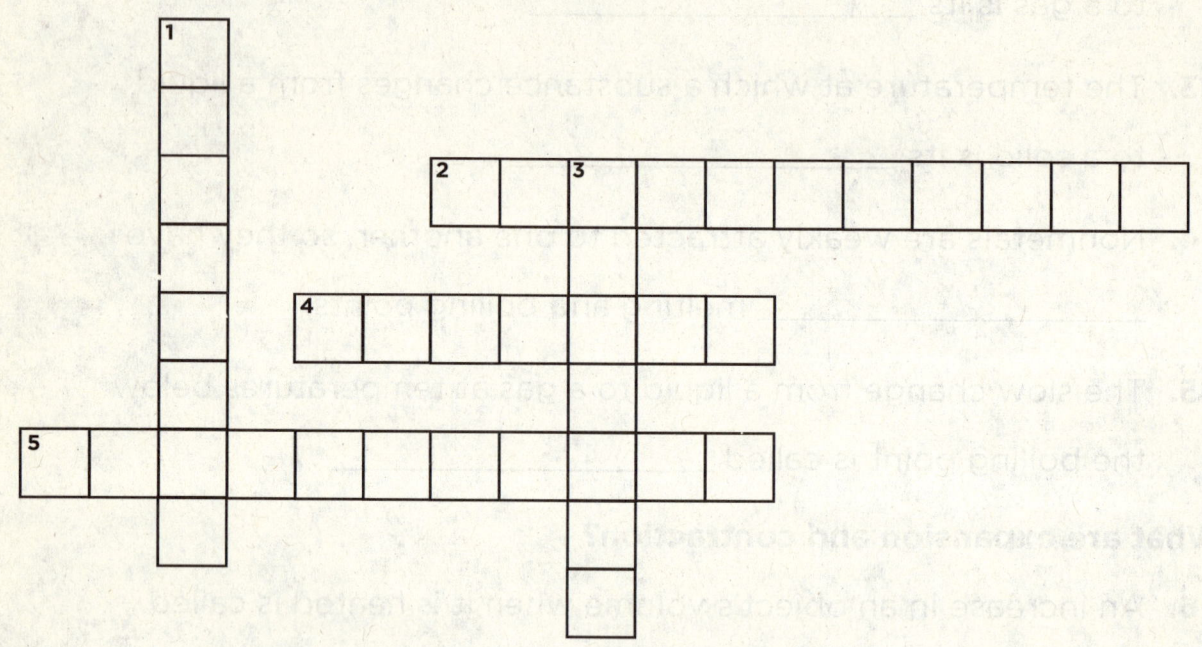

Across

2. A change from a solid to a gas.
4. The temperature at which water changes from a solid to a liquid is the _____ point.
5. A decrease in an object's volume because of a change in temperature is thermal _____ .

Down

1. The temperature at which water changes from a liquid to a solid is its _____ point.
3. The temperature at which water changes from a liquid to a gas is the _____ point.

218 Chapter 10 • Physical and Chemical Changes
Reading and Writing

Use with **Lesson 1**
Changes of State

Changes of State

Fill in the blanks.

boiling point	heat energy	solid
freezing point	liquid	sublimation
gas	melting point	temperature

All substances have three common forms called physical states. These states are _____, liquid, and _____. The physical state of matter is changed when _____ is added or taken away. A measure of the average heat energy that a substance has (the average vibration of its molecules) is its _____. When a solid is heated to its _____, its molecules start moving faster, and the solid changes into a(n) _____. When the liquid is heated to its _____, its molecules move even faster, and the liquid turns into a gas. The melting point of water is 0°C, and its boiling point is 100°C. Sometimes a solid changes directly into a gas without passing through the liquid state, a process called _____. When a liquid is cooled to its _____, it becomes a solid. When a gas is cooled, it condenses and becomes a liquid.

LESSON Outline

Name _____ Date _____

Mixtures

Use your textbook to help you fill in the blanks.

What are mixtures?

1. A physical combination of substances that remain the same is a(n) _____.

2. Mixtures can be _____ into their original substances.

3. Mixtures with different parts that can be plainly seen with the naked eye are called _____ mixtures.

4. Mixtures that look smooth to the naked eye but speckled under a microscope are called _____.

5. Over time, one or more parts of a suspension can _____.

6. A heterogeneous mixture with parts that do not settle out is called a(n) _____.

What are solutions?

7. A mixture that looks the same everywhere, even under a microscope, is called a(n) _____.

8. The part of a solution that is dissolved is the _____.

9. The part of a solution that dissolves the other substance is called the _____.

Name _____ Date _____

LESSON Outline

10. A solution of two or more solids is a(n) _____.

11. Because it can dissolve many things, water is called the _____.

How can you take mixtures apart?

12. To separate one part of a mixture from another, you can use a(n) _____.

13. When two liquids in a mixture have different boiling points, they can be separated by _____.

14. Because liquids travel at different speeds through an absorbent paper, they can be separated by _____.

How are mixtures used?

15. Cheese, gelatin, marshmallows, and paint are all examples of useful _____.

16. Copper is alloyed with zinc to make _____.

Critical Thinking

17. Suppose you were to mix together salt, water, and mud. Identify the type of mixture you have made. Describe how you could separate the parts of the mixture from one another.

_____
_____
_____
_____
_____

Chapter 10 • Physical and Chemical Changes
Reading and Writing

Use with Lesson 2
Mixtures

LESSON Vocabulary

Name _____ Date _____

Mixtures

Who am I? What am I?

Choose a word from the word box below that answers each question.

| **a.** alloy | **c.** distillation | **e.** solubility | **g.** solution |
| **b.** colloid | **d.** mixture | **f.** solute | **h.** solvent |

1. _____ I am smoke, cheese, and foam. I am a mixture that does not settle. Who am I?

2. _____ I am the water in sugar water. Who am I?

3. _____ I am the sugar in sugar water. Who am I?

4. _____ I am steel and I am brass. Who am I?

5. _____ I am the maximum amount of solute that can go into a solvent. What am I?

6. _____ Using evaporation and condensation, I can separate the liquids in a mixture. What am I?

7. _____ I am a combination of two or more materials, but none of my parts are chemically combined. What am I?

8. _____ I can be made with solids, liquids, and gases. All my parts blend so that I look the same everywhere, even under a microscope. Who am I?

Mixtures

Fill in the blanks.

alloys	distillation	liquids
boiling points	heterogeneous	solids
condensing	homogeneous	suspensions

Several substances that are physically mixed together but not chemically combined are called mixtures. Mixtures can include various combinations of solids, liquids, and gases. Liquids in a mixture may have different _____ . Boiling and _____ the liquids, a process called _____ , can be used to separate them.

There are two kinds of mixtures: those that are the same throughout (_____) and those that are not (_____). Homogeneous mixtures, such as sugar water, are called solutions. Gases form solutions more easily than _____ do, and liquids form solutions more easily than _____ do. Solutions of two or more solids are called _____ .

The different parts of some heterogeneous mixtures can clearly be seen by the naked eye. These are called _____ . Some suspensions settle to the bottom.

LESSON Outline

Name _____ Date _____

Compounds and Chemical Changes

Use your textbook to help you fill in the blanks.

What are compounds?

1. A combination of two or more elements is called a(n) _____ .

2. A compound has different properties than do the _____ that formed it.

3. Rust is a combination of iron and _____ .

4. The chemical name for rust is _____ .

5. The chemical formula for rust is _____ .

What are chemical changes?

6. Changing one substance into another is a(n) _____ .

7. When atoms break their old links and form new links with other atoms, a(n) _____ has occurred.

8. Chemists keep track of which substances are used and created in a chemical reaction by writing _____ .

9. Chemicals on the left side of a chemical equation are called _____ ; chemicals on the right side are called _____ .

10. In every chemical reaction, the total mass of the reactants always equals the total mass of the products. This fact is known as the _____ .

Chapter 10 • Physical and Chemical Changes
Reading and Writing

Use with **Lesson 3**
Compounds and Chemical Changes

Name _____ Date _____

LESSON Outline

How can you spot a chemical change?

11. A color change on metal that is caused by a chemical change is called _____.

12. Bubbles form when baking soda and vinegar are mixed, indicating that a _____ has taken place.

13. A solid that forms when two solutions are mixed is called a(n) _____.

14. If a chemical reaction produces heat and light, then reversing the reaction should _____.

How can you use chemical changes?

15. Plants use a chemical reaction called _____ to produce sugars from sunlight, water, and carbon dioxide.

16. Plants and animals use a chemical reaction called _____ to burn sugars for energy.

17. Chemical reactions are used to produce a variety of products, such as _____.

Critical Thinking

18. Write the equation for the chemical change that produces water from two hydrogen molecules and one oxygen molecule. Label the reactants and the products. (Hint: Remember to take into account the conservation of mass.)

_____
_____
_____

Chapter 10 • Physical and Chemical Changes
Reading and Writing

Use with **Lesson 3**
Compounds and Chemical Changes

Compounds and Chemical Changes

Use the words in the word box to fill in the blanks.

chemical	photosynthesis	reactants
compound	precipitate	tarnish
equations	products	

1. The _____ are on the left side of a chemical equation.

2. The _____ are on the right side of a chemical equation.

3. The chemical reaction that plants use to produce sugar is known as _____.

4. A solid that is a product of a chemical reaction is called a(n) _____.

5. Atoms break their old links and form new links during a(n) _____ change.

6. Chemists keep track of chemical reactions by using chemical _____.

7. A color change in metal caused by a chemical change is called _____.

8. A chemical combination of two or more elements is a(n) _____.

Compounds and Chemical Changes

Fill in the blanks.

> chemical equations $C_6H_{12}O_6$ H_2O
> chemical formulas   compounds   left
> CO_2 elements

    A chemical change results in one or more products that are different from the reactants. Atoms break their links and form new links with other atoms to form new _____. Chemists describe what goes on in a chemical change by writing _____. The substances to the _____ of the arrow in a chemical equation are the reactants; the substances to the right of the arrow are the products. The compounds in a chemical equation are written as _____. A chemical formula tells which _____ are in a compound and how many atoms there are of each. For example, the chemical formula for water is _____, and the chemical formula for carbon dioxide is _____. The chemical equation for photosynthesis is $6H_2O + 6CO_2 \longrightarrow$ _____ $+ 6O_2$. The numbers of atoms of each element are the same on each side of the equation.

Writing in Science

Name _____ Date _____

The Case of the Mystery Compounds

Write About It

Do research and write a report about how scientists can test water for pollutants and dangerous chemical compounds. Which chemical reactions do they use to perform the test? Give the steps of the process in order.

Getting Ideas

As you do research on how scientists test water, fill out the chart below. Write the steps in order.

First

↓

Next

↓

Last

Planning and Organizing

Organize the steps that Sean wrote about testing water for chlorine.

1. Chlorine will turn the litmus paper red, then white. _____

2. Place a sample of the water in a test tube. _____

3. Dip blue litmus paper in the water. _____

228 Chapter 10 • Physical and Chemical Changes
Reading and Writing

Use with **Lesson 3**
Compounds and Chemical Changes

Name _____ Date _____

Writing in Science

Drafting

Write a sentence to begin your report. Tell an important idea about testing water for pollutants and dangerous chemical compounds.

_____

_____

_____

Now write your report. Use a separate piece of paper. Begin with the sentence you wrote above. Then tell the steps scientists follow to test water. Be sure to include important facts and details about chemical reactions.

Revising and Proofreading

Here are some sentences Sean wrote. They are very wordy. Read each pair. Combine each pair into one sentence by cutting out unnecessary words. Write the new sentence on the line.

1. Make sure the test tube you use is clean. It must be sterile.

_____

2. The chemical reaction may produce changes in color. It may produce changes in smell.

_____

3. Test the sample quickly. Do the test within two hours.

_____

Now revise and proofread your writing. Ask yourself:
▶ Did I tell the steps of testing water in order?
▶ Did I explain the chemical processes involved?
▶ Did I correct all errors?

Chapter 10 • Physical and Chemical Changes
Reading and Writing

Use with **Lesson 3**
Compounds and Chemical Changes **229**

LESSON Outline Name _____ Date _____

Acids, Bases, and Salts

Use your textbook to help you fill in the blanks.

What are acids and bases?

1. A substance that tastes _____, turns blue litmus to red, and reacts with metals to make _____ is a(n) _____.

2. When acids dissolve in water, they release _____.

3. An atom or a molecule that has lost or gained one or more electrons is a(n) _____.

4. Hydrogen ions have a positive charge because they have lost an _____.

5. Our stomachs produce _____, which helps digest food.

6. A substance that tastes _____, is slippery to the touch, and turns red litmus to blue is a(n) _____.

7. When bases dissolve in water, they release _____, which have a(n) _____ charge.

8. _____ is used to make fertilizers.

9. Sodium hydroxide (NaOH), also called _____, is used to make textiles, detergents, and some plastics.

Name _____ Date _____

LESSON Outline

How can indicators identify acids and bases?

10. A dye that reacts chemically with acids and bases to produce one color in acids and another color in bases is called a(n) _____.

11. A low number on the pH scale indicates _____; a high number indicates _____.

12. A pH of 7 means that the solution is _____.

What are salts?

13. Mixing an acid with a base produces _____ and water.

14. Acids and bases combine to form pH neutral solutions, a process called _____.

15. A compound that has positive and negative ions in a regular pattern or crystal is a(n) _____.

16. Acids, bases, and salts dissolve in water to form a(n) _____.

Critical Thinking

17. Compare and contrast acids and bases. Tell what happens when they are mixed together.

_____
_____
_____
_____

Chapter 10 • Physical and Chemical Changes
Reading and Writing

Use with **Lesson 4**
Acids, Bases, and Salts

LESSON Vocabulary

Acids, Bases, and Salts

Who am I? What am I?

Choose a word from the word box below that answers each question.

| a. acid | c. alkalinity | e. electrolyte | g. neutralization |
| b. acidity | d. base | f. ion | h. pH |

1. _____ I can dissolve in water to form ions, which allows me to conduct electricity. Who am I?

2. _____ I have lost or gained electrons, which gives me a positive or negative charge. Who am I?

3. _____ I represent the strength of an acid. What am I?

4. _____ I taste sour and turn blue litmus red. In water I produce H^+ ions. Who am I?

5. _____ I can tell you how acidic or basic a substance is. What am I?

6. _____ I am the strength of a base. What am I?

7. _____ I taste bitter and feel soapy. In water I produce OH^- ions. Who am I?

8. _____ I can occur when acids and bases are mixed together. What am I?

Acids, Bases, and Salts

Fill in the blanks.

acid-base indicator	bitter	pH scale
acidity	blue	neutralize
alkalinity	high	
bases	low	

Compounds that give off hydrogen ions (H^+) when dissolved in water are called acids. They taste sour, sting to the touch, and turn red an _____ called litmus. Compounds that give off hydroxide ions (OH^-) when dissolved in water are called _____ . They usually taste _____ , feel soapy, and turn litmus _____ . The _____ measures the strength of an acid (known as _____) and the strength of a base (known as _____). Highly acidic solutions have a(n) _____ pH; very alkaline solutions have a(n) _____ pH. When acids and bases are mixed together, they produce a salt and water. Acids and bases _____ each other. The process in which an acid and a base combine to form a pH-neutral solution is called neutralization.

Reading in Science

Name _____ Date _____

Meet Christina Elson

Read the Reading in Science feature in your textbook.

Infer

Fill in the Infer graphic organizer below. Use the clues and what you know to draw conclusions about Aztec artifacts.

Clues	What I Know	What I Infer
Large pots have been found with salt crystal residue in them.	Aztecs had to boil salty water to get salt crystals.	
In one Aztec town, thousands of fragments of clay pots were found.	Salt was sold and transported in this Aztec town.	
Salt helps pigment "cling" to cloth.	Cloth was dyed with pigment in a hot, watery dye bath.	

Name _____ Date _____

Reading in Science

Write About It

Infer

1. How did the Aztecs change a mineral resource into a finished product?

2. What would happen to the colors in Aztec cloth when washed if salt were not part of the dye-bath?

What I Know

Fill in the blanks to complete each of the steps in the salt-making process. Use clues from the reading passage. Then answer the questions that follow.

a. Salt deposits are found in dried _____.

b. Salty _____ is collected by _____.

c. Then, _____ is filtered through the _____ and collected in large _____.

d. Finally, the water in the large pots is _____; it _____, leaving behind salts.

1. How did the Aztecs change a mineral resource into a finished process?

_____
_____
_____

2. What would happen to the colors in Aztec cloth if salt were not part of the dye-bath?

_____
_____
_____

Chapter 10 • Physical and Chemical Changes
Reading and Writing

Use with **Lesson 4**
Acids, Bases, and Salts

CHAPTER Vocabulary

Physical and Chemical Changes

Choose the letter of the best answer.

1. Which of the following is a physical change?
   a. paper burning
   b. egg frying
   c. water boiling
   d. baking soda and vinegar fizzing

2. Snow changing to water vapor is an example of
   a. sublimation.
   b. boiling.
   c. melting.
   d. thermal contraction.

3. When most liquids freeze, they undergo
   a. thermal expansion.
   b. thermal contraction.
   c. condensation.
   d. sublimation.

4. When a gas loses heat, it
   a. evaporates.
   b. boils.
   c. sublimates.
   d. condenses.

5. The temperature at which alcohol changes to a gas is its
   a. sublimation point.
   b. freezing point.
   c. boiling point.
   d. melting point.

6. Steel is an example of a(n)
   a. alloy.
   b. colloid.
   c. heterogeneous mixture.
   d. suspension.

7. Which of the following can form a solution most easily?
   a. two liquids
   b. two gases
   c. two solids
   d. a gas and a liquid

8. In a saltwater solution, the salt is a(n)
   a. alloy.
   b. colloid.
   c. solvent.
   d. solute.

236 Chapter 10 • Physical and Chemical Changes
Reading and Writing

Name _____ Date _____

CHAPTER
Vocabulary

9. Which of the following is an example of a colloid?

   a. gelatin

   b. brass

   c. sugar water

   d. orange juice

10. Which of the following is a compound?

    a. brass
    b. rust
    c. iron
    d. steel

11. In the chemical reaction called photosynthesis, which of the following is a reactant?

    a. sunlight

    b. oxygen

    c. carbon dioxide

    d. sugar

12. Which of the following indicates that a chemical change has taken place?

    a. a change from a liquid to a gas

    b. an increase in the volume of a substance

    c. a change from a solid to a liquid

    d. a change in the color of a substance

13. Which of the following is a property of bases?

    a. tastes bitter

    b. tastes sour

    c. stings the skin

    d. reacts with metal to make hydrogen gas

14. Which of the following releases hydrogen ions when dissolved in water?

    a. sodium hydroxide

    b. hydrochloric acid

    c. sodium chloride

    d. baking soda

15. What happens when an acid and a base are mixed?

    a. A gas is given off.

    b. A salt forms.

    c. A color change occurs.

    d. Heat is given off.

Chapter 10 • Physical and Chemical Changes
Reading and Writing

UNIT Literature

The Great Jump in China

Read the Literature feature in your textbook.

Write About It

Response to Literature This article describes how an athlete used a ramp to jump over a large object. If you were a professional athlete, what other kinds of devices might you use? Write a fictional narrative describing your device and its uses.

Name _____ Date _____

CHAPTER
Concept Map

Using Forces

Fill in the concept map below using the information you know about forces.

---

1. Motion is a change in an object's _____ over time.

↓

2. Speed is a measure of how fast an object's position changes. A measurement of an object's speed and its direction is _____ . A change in an object's velocity is _____ .

↓

3. A force is a push or a _____ exerted on an object.

↓

4. Newton's laws describe how forces affect _____ . These laws include the _____ , second, and _____ .

↓

5. A force multiplied by the distance over which the force is applied is _____ . The ability to do work is _____ .

↓

6. Machines can make doing work easier by changing the _____ of a force or the _____ over which the force is applied.

Chapter 11 • Using Forces
Reading and Writing

LESSON Outline

Name _____ Date _____

Motion

Use your textbook to help you fill in the blanks.

What is motion?

1. The location of an object is its _____.
   A change in the position of an object over time is motion.

   Motion has two parts: _____ and _____.

2. Distance can be measured in _____, _____, _____, or _____.

3. To measure direction, you can use a(n) _____ and units of _____.

4. You need a(n) _____ from which to measure position or motion.

What is speed?

5. To calculate speed, divide the _____ by the _____.

6. Units of speed can be _____ or _____.

7. To state the velocity of an object, you need to know the object's _____ and its _____.

Chapter 11 • Using Forces
Reading and Writing

Use with **Lesson 1**
Motion

Name _____ Date _____

LESSON Outline

What is acceleration?

9. Any change in the velocity of an object is a(n) _____.

10. If the speed of a car traveling south is increasing 5 m/s every second, its acceleration is _____.

11. An acceleration can be a change in speed or a change in _____. Negative acceleration is called _____.

What is momentum?

12. An object's mass multiplied by its velocity is its _____.

13. An object with a mass of 1 kg and a velocity of 10 m/s has a momentum of _____.

14. The more mass an object has, the _____ its inertia.

Critical Thinking

15. Would it be more difficult to stop a truck carrying a heavy load or stop the same truck empty? Explain your answer, using the concepts of inertia and momentum.

_____
_____
_____
_____
_____

Chapter 11 • Using Forces
Reading and Writing

Use with Lesson 1
Motion

LESSON Vocabulary

Name _____ Date _____

Motion

Use the words in the word box to finish the puzzle.

| acceleration | momentum | position | speed |
| inertia | motion | reference | velocity |

Down
1. location of an object
3. change in velocity over time
6. the rate at which an object's position is changing over time
7. any change in position

Across
2. tendency of an object to resist a change in motion
4. measurement of an object's speed and direction of motion
5. a *"frame"* from which you can measure position or motion
8. mass times velocity

242 Chapter 11 • Using Forces
Reading and Writing

Use with **Lesson 1**
Motion

Motion

Fill in the blanks.

| acceleration | motion | time |
| momentum | speed | velocity |

To describe how an object moves, you need a frame of reference, or a group of objects from which you can measure position. You can then measure the object's _____, or change in position. By dividing the distance an object moved by the _____ it took to move that distance, you describe an object's average _____. If you also measure the direction in which the object moved, you can describe its _____. If you know an object's instantaneous speed at the beginning and end of a time interval, you can describe the object's _____ over that time interval.

An object's mass multiplied by its velocity is its _____. The greater an object's inertia or resistance to a change in its motion, the greater its momentum.

Reading in Science

Name _____ Date _____

The Position of Earth and the Sun

Read the Reading in Science feature in your textbook.

Main Idea and Details

Use the table below to record the main idea and details described in the timeline portion of the reading passage in your textbook.

Main Idea	Details
Many throughout history have made discoveries that help us determine how the planets and stars move.	Aristotle developed a model showing the _____ around _____ .
	Ptolemy used Aristotle's model and _____ to predict the way the Sun, the Moon, and planets would appear in the _____ .
	_____ first proposed that the Sun is at the center of the Solar System.
	Galileo's discovery of _____ circling _____ supported Copernicus's theory.
	Einstein explained how _____ works, helping us understand the movement of planets and stars.
	_____ worked on the first 3-D map of the _____ .

244 Chapter 11 • Using Forces
Reading and Writing

Use with Lesson 1
Motion

Name _____ Date _____

Reading in Science

Write About It

Main Idea and Details Read the "Write About It" question. Use the text of "The Position of Earth and the Sun" feature to write your answers.

Identifying the Main Idea

The main idea is the central point of the passage. It tells you what the passage is about. Review the graphic organizer to find the main idea of the passage. Write that idea on the lines below.

_____
_____
_____

Identifying Supporting Details

Details are important parts of the passage that support the main idea. Look for the supporting details within the list of scientists that follows the opening paragraphs. Give one detail from the article that supports the main idea. You can choose one supporting detail from your table.

_____
_____
_____
_____

Chapter 11 • Using Forces
Reading and Writing

Use with **Lesson 1** Motion   **245**

LESSON Outline   Name _____   Date _____

Forces and Motion

Use your textbook to help you fill in the blanks.

What are forces?

1. Units of force are the _____ and the _____ .

2. An arrow can be used to represent the _____ and _____ of a force.

3. Three forces that act on an airplane: _____ , lift, and _____ .

What are gravity and friction?

4. The force that pulls all objects together is called _____ .

5. The amount of friction depends on two factors: the roughness of the _____ of the objects and how much force is required to _____ the two objects together.

6. _____ is created whenever there is friction.

What is Newton's first law?

7. According to the law of inertia, an object at rest tends to _____ , and an object in motion tends to _____ , unless acted upon by an _____ .

Name _____ Date _____

LESSON Outline

What is Newton's second law?

8. According to Newton's second law, an object's acceleration increases as the amount of unbalanced force on it _____ ; an object's acceleration decreases as the object's mass _____ .

What is Newton's third law?

9. When one object pushes on a second object, the second object pushes back on the first object with the same amount of _____ .

10. According to Newton's third law, for every action there is a(n) _____ but _____ reaction.

Critical Thinking

11. Suppose that you are walking down the street. Describe the forces acting on you, and use Newton's laws of motion to describe your motion.

_____
_____
_____
_____
_____
_____
_____
_____

Chapter 11 • Using Forces
Reading and Writing

Use with **Lesson 2**
Forces and Motion **247**

LESSON Vocabulary

Name _____ Date _____

Forces and Motion

What am I?

Choose a word from the word box below that answers each question.

a. action force	**d.** friction	**g.** unbalanced
b. balanced	**e.** inertia	
c. force	**f.** reaction force	

1. _____ I am the word that scientists use for a push or a pull. What am I?

2. _____ I am the force that sometimes makes sliding difficult. What am I?

3. _____ I am a force whose effect is offset by other forces, so I won't change your motion. What type of force am I?

4. _____ I am a force whose effect is not offset, so I change your motion in some way. What type of force am I?

5. _____ I am the first force in a pair. Whatever I push pushes back on whatever caused me. What am I?

6. _____ I am the second force in a pair. If something gets pushed, I push back. What am I?

7. _____ I am the tendency of an object in motion to stay in motion.

248 Chapter 11 • Using Forces
Reading and Writing

Use with **Lesson 2**
Forces and Motion

Forces and Motion

Fill in the blanks.

| accelerate | force | gravity | mass |
| distance | gravitation | inertia | unbalanced |

The motion of any object can be explained using the laws that Newton discovered more than 300 years ago. His universal law of _____ states that objects with more _____ have more force of _____ between them. Objects that are separated by more _____ have less force of gravity between them.

According to Newton's first law, also called the law of _____ , an object at rest tends to stay at rest, and an object in motion tends to stay in motion, unless acted upon by a(n) _____ force. The second law can be summed up with the equation F = ma. This equation means that an object accelerates more as the size of the unbalanced _____ on it increases and that more massive objects _____ less for a given force. Newton's third law states that for every action force there is an equal and opposite reaction force.

Chapter 11 • Using Forces
Reading and Writing

LESSON Outline

Name _____ Date _____

Work and Energy

Use your textbook to help you fill in the blanks.

What is work?

1. Work done on an object changes the amount of _____ that the object has.

2. Work is equal to the _____ used multiplied by the _____ over which the force was applied.

3. The units of work are _____ , or _____ .

4. Work occurs when _____ cause an object to accelerate.

5. Total work is the sum of _____ work and _____ work.

6. When you move an object, _____ often performs negative work on it.

What is energy?

7. Energy is measured in units called _____ .

8. A stretched spring has _____ energy.

   A moving object has _____ energy.

9. Doing positive work on an object increases its _____ .

10. Throwing a ball increases its _____ energy; lifting a ball increases its _____ energy.

250 Chapter 11 • Using Forces
Reading and Writing

Use with **Lesson 3**
Work and Energy

Name _____ Date _____

LESSON Outline

11. Chemical energy, nuclear energy, and magnetic energy are different forms of _____ energy.

12. Heat, electricity, sound, and light are different forms of _____ energy.

How can energy change?

13. Energy cannot be _____ or _____ ; it can only _____ .

14. Whenever energy is used to do work, energy _____ .

15. Kinetic energy is often changed into heat energy by _____ .

Critical Thinking

16. Trace the energy changes that occur in a toaster, in a radio, and in a windmill used to generate electricity.

_____
_____
_____
_____
_____
_____
_____
_____

Chapter 11 • Using Forces
Reading and Writing

Use with **Lesson 3**
Work and Energy

Work and Energy

Use the words in the word box to finish the puzzle.

chemical	joules	sound
conservation	kinetic	
electricity	potential	

Down

1. Energy that is stored in the position of an object is called _____ energy.

2. Units of work are _____.

3. The energy of a moving object is _____ energy.

4. The kinetic energy of electrons is called _____.

5. The kinetic energy of particles as they move in waves is _____.

Across

6. The law of _____ of energy states that energy cannot be created or destroyed; it can only change form.

Work and Energy

Fill in the blanks.

| destroyed | friction | positive | sound |
| electrical | kinetic | potential | work |

Work is defined as an unbalanced force acting on an object through a certain distance. The total work done on an object is the sum of the _____ work and the negative work done on it. The force of _____ usually does negative work on a moving object. Energy is defined as the ability to do _____ .

If you lift a ball, you give it _____ energy. If you drop the ball, its potential energy is converted into _____ energy. Different forms of potential energy include chemical, nuclear, magnetic, and _____ energy. Different forms of kinetic energy include electricity, _____ , and light. The law of conservation of energy states that energy cannot be created or _____ . Energy can only change forms.

LESSON Outline

Name _____ Date _____

Simple Machines

Use your textbook to help you fill in the blanks.

What are simple machines?

1. A simple machine can change the _____, _____, or _____ of a force that you apply.

2. When you apply a force to a machine's _____ arm, the machine applies an output force to the load through its _____ arm.

3. The ratio of a machine's output force to the effort applied is called its _____.

What are levers?

4. A lever can either multiply an _____ or multiply _____.

5. A crowbar is a _____ lever—the effort arm and the _____ are on opposite sides of the _____.

6. A wheelbarrow is a _____ lever—the effort force is _____ than the output force, and both are in the same _____.

7. A fishing rod is a _____ lever—its output force is _____ than the effort force, but output distance of the tip of the rod is greater than the effort distance of your hand.

Name _____ Date _____

LESSON Outline

Which machines are like levers?

8. A wheel and axle is a type of lever in which the axle acts like the _____ and the wheel acts like the _____ of the lever.

9. A wheel and axle with a free-moving cord is called a _____.

What are inclined planes?

10. An inclined plane that is used to separate two objects is called a(n) _____. An inclined plane wrapped around a cylinder is a(n) _____.

11. The farther apart the threads of a screw, the _____ the screw moves when turned, but the _____ effort it takes to turn it.

What are compound machines?

12. Any machine that combines two or more simple machines is a _____.

13. The more work that a machine does for a given input of energy, the more _____ it is.

14. Efficiency is often expressed as a(n) _____.

Critical Thinking

15. What types of simple machines are in a wheelbarrow?

_____

_____

Chapter 11 • Using Forces
Reading and Writing

Use with Lesson 4
Simple Machines

LESSON Vocabulary   Name _____   Date _____

Simple Machines

What am I?

Choose a word from the word box below that answers each question.

compound machine	fulcrum	simple machine
efficiency	load	
effort	screw	

1. _____ I am a bicycle, car, or anything else made up of two or more simple machines. What am I?

2. _____ I am the push on a lever or the pull on a pulley. I am any force that you apply to a machine. What am I?

3. _____ I take one force and change it into another force. I can change the direction, strength, or distance of a force. What am I?

4. _____ When the effort arm goes down, the resistance arm goes up, but I don't move. I am the pivot point on a lever. What am I?

5. _____ When you push down on a lever, I am the object moved by the resistance arm. What am I?

6. _____ I can tell you how much you can gain by using a machine. I am the ratio of your input energy to the machine's output work. What am I?

7. _____ I am an inclined plane wrapped around a cylinder. What am I?

Simple Machines

Fill in the blanks.

| farther | less | longer | simple machine |
| fulcrum | load | resistance | strength |

Simple machines make work easier by changing the distance, direction, or amount of the effort force that you apply. Using an inclined plane, you can raise an object with less effort than if you lifted it directly upward. The _____ the inclined plane, the less effort needed to lift a load. A pulley can change the direction or _____ of the force applied to lift a load. A lever has an effort arm, resistance arm, and _____, or pivot point. When you apply a force on the effort arm, the _____ arm applies a force on the _____. If the effort arm is longer than the resistance arm, you use _____ force to lift a load, but the effort arm moves _____.

Compound machines combine two or more _____. The more work a machine does for a given input of energy, the more efficient the machine is.

Writing in Science

Name _____ Date _____

A Humane Mousetrap

Write About It

Do some online research about bird feeders that keep squirrels from stealing the birdseed. Write an explanation of how this kind of bird feeder works by using simple machines. Provide steps for making this device. (You can invent your own.)

Getting Ideas

Do some online research on birdfeeders. Then fill in the sequence chart below. Jot down steps for making a birdfeeder that keeps squirrels from stealing the birdseed.

First
↓
Next
↓
Then
↓
Finally

Planning and Organizing

When organizing explanatory writing, it is often best to write the details as they happened. Write the detail that happened first. Then the detail that happened second. Then the detail that happened last. When writing your explanation, make sure you write your steps in the order they happen.

Name _____ Date _____

Writing in Science

Drafting

Write a sentence to begin your explanation. Tell what your birdfeeder does. In other words, tell how it is squirrel-proof.

_____
_____

Now write your explanation. Use a separate piece of paper. Begin with the sentence you just wrote. Tell how the birdfeeder works. Then tell the steps for making it. Write these steps in time order.

Revising and Proofreading

Here is part of Alicia's explanation. Combine each pair of sentences. Use the word in parentheses.

1. Squirrels slide down the pole. It is slippery. (because)
   _____

2. Squirrels can't jump onto the top of the feeder. It is too high up. (since)
   _____

3. Birds can get at the seeds. There are holes in the mesh. (because)
   _____

4. Do not put this feeder under a tree. A squirrel might jump down onto it. (since)
   _____
   _____

Now revise and proofread your writing. Ask yourself:

▶ Did I clearly and accurately explain how the birdfeeder works?
▶ Did I write the steps for making it in order?
▶ Did I correct all mistakes?

Chapter 11 • Using Forces
Reading and Writing

Use with **Lesson 4**
Simple Machines

Using Forces

Choose the letter of the best answer.

1. How fast an object's position is changing over time is the object's
   a. velocity.
   b. acceleration.
   c. speed.
   d. mass.

2. Momentum is calculated by multiplying an object's mass by its
   a. mass.
   b. velocity.
   c. work.
   d. inertia.

3. The force of gravity between two objects
   a. increases with mass and decreases with distance.
   b. increases with distance and decreases with mass.
   c. decreases with mass and increases with distance.
   d. increases with mass and increases with distance.

4. Friction between objects produces
   a. gravity.
   b. load.
   c. inertia.
   d. heat.

5. Newton's second law of motion states that force is equal to mass times
   a. speed.
   b. energy.
   c. velocity.
   d. acceleration.

6. Placing a dish on a higher shelf increases the dish's
   a. inertia.
   b. kinetic energy.
   c. weight.
   d. potential energy.

Name _____ Date _____

CHAPTER Vocabulary

Choose the letter of the best answer.

7. Work is done when
   a. you push against a wall.
   b. you lift a book.
   c. you stand on the floor.
   d. you hold a box.

8. When you do positive work on an object, you
   a. decrease the object's energy.
   b. keep the object's energy the same.
   c. increase the object's energy.
   d. may increase or decrease the object's energy.

9. The unit that is used to measure force is the
   a. meter.
   b. kilogram.
   c. Newton.
   d. joule.

10. The force that you apply to a simple machine is called the
    a. effort.
    b. work.
    c. load.
    d. output.

11. If a machine is 50 percent efficient, how much energy must you apply to lift a 100-Newton weight a distance of 10 meters?
    a. 2000 joules
    b. 1000 joules
    c. 500 joules
    d. 100 joules

12. Which of the following is an example of an inclined plane?
    a. pulley
    b. ramp
    c. gear
    d. wheel and axle

13. Which of these is a compound machine?
    a. wedge
    b. screw
    c. pair of scissors
    d. wheel and axle

Chapter 11 • Using Forces
Reading and Writing

CHAPTER Concept Map

Name _____ Date _____

Using Energy

Fill in the concept map below, using information you know about energy.

	Definition	Example
Heat	Heat is energy that flows because of a difference in _____.	The energy that flows away from your hand when you hold a _____ drink
Sound	Sound is energy that moves in the form of a _____ that is a series of compressions and _____.	The energy from a whistle is an example of sound that has a high _____.
Light	Light is a wave made from electric and _____ energy. Light is also a _____.	The light from a rainbow is an example of light that is spread out into a _____.
Electricity	Electricity is energy in the form of moving _____.	One example of electricity is the movement of _____ that occurs when you touch a door knob.
Magnetism	Magnetism is the ability of one object to _____ or _____ on another object that has the same magnetic property.	Magnetism is shown when two magnets either _____ or _____ each other.

262 Chapter 12 • Using Energy
Reading and Writing

Name _____ Date _____

LESSON Outline

Heat

Use your textbook to help you fill in the blanks.

What is heat?

1. Heat is energy that moves from an object with a(n) _____ temperature to an object with a(n) _____ temperature.

2. Heat continues to flow from one object to another object until both have the same _____.

3. Heat is the _____ amount of thermal energy that an object releases.

How does heat travel?

4. Conduction can occur between objects that are _____.

5. As hot and cool portions of a liquid or gas move, _____ currents form.

6. The heat that you can feel radiating away from hot objects as electromagnetic rays is called _____ rays.

What is thermal conductivity?

7. Convection currents move heat more slowly than do _____ but more quickly than conduction.

8. Heat traveling by conduction moves at the speed at which molecules can _____ one another and change how fast nearby molecules are vibrating.

LESSON Outline

Name _____ Date _____

9. A material that conducts heat poorly is a good

   _____.

10. Thermal conductivity increases as _____

    increases, so _____ are the best

    conductors of heat and _____ are the
    worst conductors.

11. Objects with a low heat capacity change temperature

    _____ when heated.

When is heat waste?

12. Heat energy caused by friction is usually a waste product

    that results when energy _____ or

    _____.

Critical Thinking

13. Describe how heat is used in a kitchen. What appliances produce heat, and how do they produce it? What objects are used as insulators, and what objects are used as conductors?

    _____
    _____
    _____
    _____
    _____
    _____
    _____

Name _____ Date _____

LESSON Vocabulary

Heat

Who am I? What am I?

Choose a word from the word box below that answers each question.

a. conduction	d. heat
b. conductivity	e. radiation
c. convection	f. temperature

1. _____ I can transfer heat through a vacuum because I am electromagnetic rays. Who am I?

2. _____ I flow from a warmer object to a cooler object until both objects are the same temperature. What am I?

3. _____ I move heat through a material from one atom or molecule to the next. Who am I?

4. _____ I move heat as a liquid or a gas rises and sinks. Who am I?

5. _____ I am a measurement of the average energy of molecules. What am I?

6. _____ I can tell you how easily heat moves through a material. What am I?

Chapter 12 • Using Energy
Reading and Writing

Use with Lesson 1
Heat

LESSON Cloze Activity

Heat

Fill in the blanks.

conduction	gases	temperature
convection	liquids	thermal conductors
faster	molecules	thermal insulators

Heat is energy that flows from an object at a higher temperature to an object at a lower temperature. The measure of the average kinetic energy of molecules is _____ . When a warmer object touches a cooler object, heat moves by _____ . The molecules of the warmer object vibrate _____ . The two objects stay in place, but their _____ bump one another and energy passes from the warmer object to the cooler object.

Some materials, such as metals, are good _____ . Other materials, such as gases, are good _____ . Currents of matter spread heat through _____ and _____ , a process called _____ . The transfer of heat by electromagnetic rays is called radiation.

Name _____ Date _____

LESSON Outline

Sound

Use your textbook to help you fill in the blanks.

How is sound produced?

1. Regions of a material that have many molecules squeezed together are _____ ; regions that have fewer molecules spread apart are _____ .

2. A series of compressions and rarefactions moving through a medium is a(n) _____ .

3. Sound waves vibrate the medium in the _____ direction that the energy moves.

How does sound travel?

4. Sound cannot travel through a(n) _____ , which is a region of space that contains no matter.

5. Sound travels faster through a(n) _____ than it travels through a liquid or a(n) _____ .

6. When sound hits soft, thick, or uneven materials, much of the sound is _____ ; when sound hits flat, firm surfaces, much of it is _____ .

What is pitch?

7. The higher the frequency or pitch of a sound wave, the more _____ pass in a period of time.

8. To increase the pitch of a musical instrument, you need to _____ the part that vibrates.

Chapter 12 • Using Energy
Reading and Writing

Use with **Lesson 2** Sound

9. If you move in the direction from which a sound wave is coming, you hear a higher pitch as a result of the _____ effect.

What is volume?

10. Amplitude of sound depends on how _____ the air in compressions is compared to normal air.

11. Volume is measured in _____ .

12. A 30 dB noise has _____ more energy than a 10 dB noise, but a 30 dB noise sounds about _____ as loud as a 10 dB noise.

13. To make a sound louder, you need to use more energy, which increases the _____ of the particles in the compressions.

14. The volume of a sound decreases with _____ because the same amount of sound energy is spread over a larger and larger area.

What is echolocation?

15. Bats make sound and listen to the _____ to locate prey.

16. Sound navigation and ranging, or _____ , is used to find the depth of a body of water and locate objects beneath water.

Critical Thinking

17. Why is the pitch of a train's whistle higher as the train approaches and lower as it moves away?

_____

_____

Name _____ Date _____

LESSON Vocabulary

Sound

Use the words in the word box to finish the sentences.

absorption	frequency	reflection
amplitude	medium	sound
echolocation	pitch	vacuum

1. _____ Material through which sound travels

2. _____ The bouncing of a sound wave off a surface

3. _____ Number of wave peaks that pass each second

4. _____ Height of a sound wave

5. _____ Finding objects by using echoes

6. _____ How high or low a sound is

7. _____ Space that contains few or no molecules

8. _____ Disappearance of a sound wave into a soft surface

9. _____ A series of rarefactions and compressions traveling through a medium

Chapter 12 • Using Energy
Reading and Writing

Sound

Fill in the blanks.

| amplitude | frequency | pitch | reflected |
| compressions | louder | rarefactions | sound wave |

As an object vibrates, it moves back and forth against the air around it. The air begins to vibrate, creating _____ where air molecules are pushed together and _____ where air molecules are farther apart. Compressions and rarefactions moving through a medium make a(n) _____. The number of compressions that pass each second is the sound wave's _____. A higher frequency sound has a higher _____.

When more energy is used to make sound, the sound has a higher _____. High-amplitude sounds are _____ than sounds having low amplitude. When sound waves hit a flat, firm surface, much of their energy is _____. When sound waves hit a soft or uneven surface, much of their energy is absorbed.

Name _____ Date _____

LESSON Outline

Light

Use your textbook to help you fill in the blanks.

What is light?

1. Light is vibrating _____ and _____ energy.

2. Light waves vibrate in directions _____ to the direction of their motion.

3. Light travels fastest in a _____.

4. The wavelength of a wave times its frequency is the _____ of the wave.

5. Light has properties of both _____ and _____.

How does light make shadows?

6. Light rays bouncing off a surface at random angles is called _____.

7. If most light goes through an object, the object is _____; if some light goes through, the object is _____; if no light goes through, the object is _____.

8. Objects that do not allow light to pass through cause _____.

Chapter 12 • Using Energy
Reading and Writing

Use with **Lesson 3** Light

LESSON Outline

Name _____ Date _____

How does light bounce and bend?

9. According to the law of reflection, the angle between an _____ light ray and a surface equals the angle between the _____ light ray and the surface.

10. When light enters a different medium, its _____ changes and it undergoes _____.

Why do we see colors?

11. White light is a mixture of many _____ that can be separated by a _____ to form a spectrum.

12. Opaque objects appear the color of the light they _____, but _____ objects appear the color of light they let pass through.

Is all light visible?

13. Many forms of _____ radiation cannot be seen with the human eye.

Critical Thinking

14. Why does mixing the primary colors of light produce white light, but mixing paints that have the primary colors produces black paint?

_____
_____
_____
_____

Name _____ Date _____

Light

Use the words in the word box to fill in the blanks.

electromagnetism	prism	translucent
image	refraction	wavelength
photon	spectrum	

1. _____ Band of colors in a rainbow

2. _____ Tiny bundle of light

3. _____ Picture of a light source that light rays make when they reflect from a mirror or refract through a lens

4. _____ The way in which electric and magnetic forces interact

5. _____ Cut piece of glass with two opposite sides in the shape of a triangle

6. _____ Distance between one peak and the next in a wave

7. _____ Material that allows only some light to pass through

8. _____ The bending of light waves as they pass from one substance to another

Light

Fill in the blanks.

colors	red	straight lines	wavelength
opaque	refracts	vacuum	
prism	spectrum	violet	

A light wave is energy in the form of electric and magnetic fields. Light travels fastest through a(n) _____ and travels slower in other mediums. The size of a light wave is measured as its _____, the distance from one peak to the next. We see different wavelengths as different _____. The shortest wavelength looks _____, and the longest wavelength looks _____.

Light travels in _____ until it strikes an object or another medium. When light enters another medium, it slows down and _____, or bends. When white light travels through a(n) _____, a triangular piece of glass, it refracts and separates into the different colors of the _____. If light strikes an _____ object, most of it is absorbed but some scatters off the object.

Name _____ Date _____

Writing in Science

How We Use Lasers

Write About It
Find out more about one of the uses of lasers. Write an expository essay giving important information about this use. Support your main idea with facts and details. Reach a conclusion at the end.

Getting Ideas
Brainstorm a list of uses of lasers. Choose one to write about. Then do some research. Use the chart below to record information that you find.

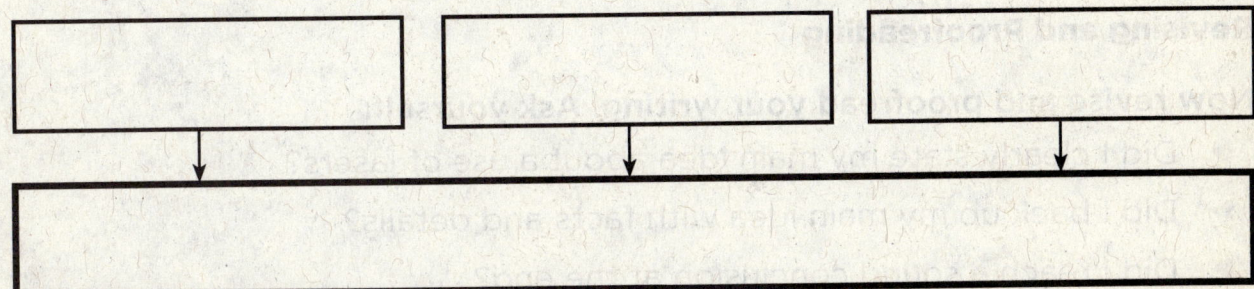

Planning and Organizing
Nick wanted to tell about the benefits of laser surgery. Here are four sentences he wrote. Write Yes if the sentence below belongs in his essay. Write No if it does not.

1. The laser seals the blood vessels when it cuts. _____

2. When lasers are used, there is less blood lost during surgery. _____

3. Scientists used lasers to measure the distance between earth and the moon. _____

4. Lasers cut down on the risk of getting an infection from surgery. _____

Chapter 12 • Using Energy
Reading and Writing

Use with **Lesson 3**
Light

Writing in Science

Name _____ Date _____

Drafting

Write a sentence to begin your essay. Tell your topic. This is the use of lasers you chose to write about. Tell your main idea about this topic. This sentence is your topic sentence.

_____

_____

_____

Now write your essay. Use a separate piece of paper. Start with your topic sentence. Then include facts and details that back up your main idea. (Do not include facts and details that don't support your main idea.) Reach a conclusion about your topic at the end.

Revising and Proofreading

Now revise and proofread your writing. Ask yourself:

▶ Did I clearly state my main idea about a use of lasers?
▶ Did I back up my main idea with facts and details?
▶ Did I reach a sound conclusion at the end?
▶ Did I correct all mistakes?

Name _____ Date _____

LESSON Outline

Electricity

Use your textbook to help you fill in the blanks.

What is static electricity?

1. When two objects rub against each other, electrons can move from one object to the other and cause a buildup of _____ electricity.

2. Electrons jumping through the air to an area that has a positive charge form a(n) _____ .

3. Charges move easily on a good _____ .

4. Objects can be protected from the buildup of static electricity by _____ them to the Earth.

How can electricity flow?

5. Circuits must have an unbroken path of conductors and a(n) _____ that causes the electrons to move along the path.

6. A device that can open or close a circuit is called a(n) _____ .

7. Resistance is measured in _____ , and electric current is measured in _____ .

8. The amount of _____ moving in a circuit is measured in units called _____ or amps (A).

Chapter 12 • Using Energy
Reading and Writing

Use with Lesson 4
Electricity

9. A current of electrons moving through resistors loses energy that changes into _____ .

What kinds of circuits are there?

10. A circuit with only one conductive path is a(n) _____ circuit; a circuit with more than one conductive path is a(n) _____ circuit.

11. In a series circuit, resistance increases with each _____ added.

12. In a parallel circuit, paths with greater _____ have less electric current flowing through them.

How can you use electricity safely?

13. To protect against large currents, homes have _____ .

14. Outlets in kitchens and bathrooms have _____ that turn an outlet off if a short is detected.

15. Touching two _____ at the same time or touching one power line and _____ or some grounded object can be deadly.

Critical Thinking

16. When a home circuit breaker opens, the lights in some rooms go off but the lights in other rooms stay on. Explain why.

_____

_____

_____

Name _____ Date _____

LESSON Vocabulary

Electricity

Matching

Match the correct letter with the description.

> a. circuit
> b. electric current
> c. grounding
> d. resistor
> e. static electricity
> f. switch

1. _____ device that opens or closes an electric circuit

2. _____ a buildup of charged particles

3. _____ an unbroken path of conductors through which electric current passes

4. _____ a conductor sharing its excess charge with a much larger conductor

5. _____ an object in an electric circuit that resists the flow of electrons

6. _____ a flow of electricity through a conductor

LESSON Cloze Activity

Name _____ Date _____

Electricity

Fill in the blanks.

amperes	resistor	voltage source
circuit	static electricity	
conductors	switch	

When objects rub against each other, electrons sometimes move from one object onto the other. The resulting buildup of charged particles is called _____. A _____ is formed when an electric current passes through an unbroken path of _____. A _____ is needed to move electrons along the circuit. The amount of electric charge moving in a circuit is measured in _____ or amps (A).

A device that opens or closes a circuit is called a(n) _____. Any device, such as a light bulb, that resists the flow of electrons is a(n) _____. Circuits that have only one path for electrons are series circuits, and circuits that have more than one path are parallel circuits.

Name _____ Date _____

Reading in Science

Building a Better Battery

Read the Reading in Science feature in your textbook.
Try to draw conclusions from text clues.

Draw Conclusions

Fill in the Drawing Conclusions Chart using text clues you find in the article.

Text Clues	Conclusion
1. Batteries are devices that store _____ and make it available in _____ form.	Batteries convert _____ energy to _____ energy.
2. All batteries have positive and negative _____ and an electrolyte through which a(n) _____ can flow.	_____ can be attached to electrodes; _____ is a solution through which electrons move.
3. A(n) _____ battery has two electrodes in a(n) _____ solution; cars today still use them.	This type of battery uses _____ electrodes, and _____ solution is the electrolyte; this type of battery can be recharged.
4. Laptop computers use _____ batteries; they are lightweight and powerful.	Lithium-ion batteries allow laptop computers to run _____ without needing to be recharged.

Chapter 12 • Using Energy
Reading and Writing

Use with **Lesson 4** Electricity

281

Reading in Science

Name _____ Date _____

Write About It
Draw Conclusions
1. What makes batteries useful?
2. What is the electrolyte in a lead acid battery?

Planning and Organizing
Answer these questions in more detail.

What things do you use that require batteries?
_____
_____

Explain what an electrolyte is.
_____
_____

Explain how a voltaic pile is constructed and what is used as the electrolyte.
_____
_____
_____

What kinds of batteries do cars have and why do they have them?
_____
_____

What are the benefits of using rechargeable batteries?
_____
_____
_____

Name _____ Date _____

LESSON Outline

Magnetism

Use your textbook to help you fill in the blanks.

What is magnetism?

1. When a magnet is cut in half, each of the two pieces has a(n) _____ pole and a(n) _____ pole.

2. Like poles of a magnet _____ each other, and unlike poles _____ each other.

3. The Earth is a giant permanent _____.

4. Whenever an electric charge moves, it creates _____ forces.

5. The _____ together the lines of a magnetic field, the stronger the magnetic force.

What are electromagnets?

6. An electric current that produces a magnetic field is called a(n) _____.

7. A magnetic field _____ around a straight wire when current is flowing through it.

8. Wrapping many loops of wire together _____ the magnetism of the coil.

9. You can increase the strength of an electromagnet in three ways: _____, place an iron rod inside the coils, or _____.

Chapter 12 • Using Energy
Reading and Writing

Use with Lesson 5
Magnetism

LESSON Outline

Name _____ Date _____

10. As the electric current rises and falls in the _____ of a speaker, its magnetic field changes, causing a cone of paper or metal to vibrate.

11. In an electric motor, a coil acting as an electromagnet rotates between the poles of a(n) _____.

How can magnets produce electricity?

12. A generator creates an electric current by spinning a coil of wire between the poles of a powerful _____.

13. The energy needed to spin the coils in an electric generator can come from _____ in a hydroelectric dam, _____ in a coal-fired power plant, or from wind or tides.

What is magnetic levitation?

14. Two electromagnets can push against each other to _____ an object.

15. Scientists have designed _____ trains that are held just above their tracks by electromagnets, greatly reducing the amount of energy lost to _____.

Critical Thinking

16. In what way is an electric generator the opposite of an electric motor?

    _____
    _____
    _____
    _____

Name _____ Date _____ **LESSON Vocabulary**

Magnetism

Who am I? What am I?

Choose a word from the word box below that answers each question.

a. alternating current	**d.** magnetic field
b. electromagnet	**e.** magnetic levitation
c. generator	**f.** magnetism

1. _____ When my wire coils are spun between the poles of my powerful magnet, I produce electricity. Who am I?

2. _____ I move back and forth through a wire, changing directions many times per second. What am I?

3. _____ I use magnetic forces to lift objects. I can even lift an entire train! Who am I?

4. _____ I am magnetic when an electric current flows through me. Who am I?

5. _____ When you have me, you can push or pull on another object that also has me. What am I?

6. _____ I describe the strength and direction of a magnet's force. If you sprinkle iron filings near a magnet, you can see me. What am I?

Chapter 12 • Using Energy
Reading and Writing

Use with **Lesson 5**
Magnetism

Magnetism

Fill in the blanks.

> electric current   magnetic field   poles
> electric motor   north   south
> electromagnet   permanent magnet   spin

Permanent magnets are made of metals such as iron.

They have two _____, north and south, and a(n)

_____ around them. An iron core with a wire

coil wrapped around it is called a(n) _____.

When a(n) _____ passes through the wire coil,

a magnetic field with a(n) _____ and a(n)

_____ pole is generated.

Electric motors and electric generators have an

electromagnet between the poles of a very strong _____.

In a(n) _____, current is sent through the wire

coil. The poles of the electromagnet switch back and forth,

causing it to _____ between poles of the

permanent magnet. In an electric generator, energy from

falling water or some other source is used to spin the wire

coil past the poles of the permanent magnet, generating

electricity in the wire coil.

Name _____ Date _____

CHAPTER Vocabulary

Using Energy

Choose the letter of the best answer.

1. A measurement of the average kinetic energy of molecules is
   a. heat.
   b. temperature.
   c. thermal capacity.
   d. thermal conductivity.

2. The movement of heat through a material while the material stays in place is
   a. radiation.
   b. convection.
   c. conduction.
   d. conductivity.

3. The surface of the Earth is warmed mainly by
   a. convection.
   b. conduction.
   c. geothermal heat.
   d. radiation.

4. Which of the following is the best thermal insulator?
   a. wood
   b. air
   c. water
   d. metal

5. Through which of these does sound travel fastest?
   a. water
   b. air
   c. metal
   d. a vacuum

6. A loud sound has a higher ____ than a soft sound.
   a. frequency
   b. pitch
   c. wavelength
   d. amplitude

7. Which statement about light is true?
   a. It has properties of both a particle and a wave.
   b. It travels slowest through a vacuum.
   c. It can travel only through matter.
   d. It always has the same amount of energy.

8. When white light travels through a prism, it forms a(n)
   a. image.
   b. spectrum.
   c. shadow.
   d. reflection.

Chapter 12 • Using Energy
Reading and Writing

287

CHAPTER Vocabulary

Name _____ Date _____

9. When light hits an opaque object, we see the color that the object
   a. absorbs.
   b. transmits.
   c. scatters.
   d. refracts.

10. What happens as light moves from one transparent material into a different transparent material?
    a. It reflects.
    b. It refracts.
    c. It is absorbed.
    d. It is scattered.

11. Which color of visible light has the shortest wavelength?
    a. red
    b. violet
    c. yellow
    d. blue

12. Which of the following is a resistor?
    a. a light bulb
    b. a switch
    c. a battery
    d. a wire

13. What happens when one bulb in a series circuit burns out?
    a. The other bulbs get dimmer.
    b. The other bulbs go out.
    c. The other bulbs get brighter.
    d. The other bulbs stay the same.

14. Which statement about magnets is true?
    a. Like poles attract each other.
    b. Unlike poles attract each other.
    c. Unlike poles repel each other.
    d. Poles have no affect on each other.

15. A device that changes electrical energy into a spinning motion is a(n)
    a. electric motor.
    b. transformer.
    c. electromagnet.
    d. electric generator.

288 Chapter 12 • Using Energy
Reading and Writing